Physics Programs

Solid State and Quantum Physics

Edited by
A. D. Boardman
Department of Pure and Applied Physics, University of Salford

A Wiley–Interscience Publication

JOHN WILEY & SONS
Chichester · New York · Brisbane · Toronto

British Library Cataloguing in Publication Data:

Physics programs.
  Solid state and quantum physics
  1. Physics—Programmed instruction
  I. Boardman, A. D.
  530'.07'7   QC21.2   80–40125

  ISBN 0 471 27734 7

Typeset in Northern Ireland, at The Universities Press (Belfast) Ltd. and printed by Pitman Press, Bath

Physics Programs

Solid State and Quantum Physics

Contents

Contents

Preface

This small book is the solid state/quantum physics section of the larger textbook called *Physics Programs* that covers the four areas, optics, magnetism, solid state/quantum physics, and applied physics. Each chapter given here is self-contained, with enough theory given for the topic discussed, and the associated computer programs, to be well understood. The programs are guaranteed in the sense that they are copied directly from fully working source texts on the computer. They can be used, possibly with minor adjustments, on any computing system. If what is required is a classroom demonstration, or the engagement of a class in a simple sequence of exercises, then the programs may be used without understanding the coding. The programs are, however, liberally strewn with comments so that they can be used for more advanced projects in which an understanding of the program is required.

The topics covered here are confined to solid state physics and quantum physics. In the first chapter a study is made of the properties of elastic waves in crystalline solids for any crystal class. In the second the familiar quantum mechanical problem of the particle in an infinite well with a perturbation is re-examined, from a computational point of view. The third chapter deals, in detail, with phonon dispersion and the phonon density of states of cubic symmetry crystals, while the final chapter is a very instructive computer study of an electron energy band problem. It is hoped that this set of chapters will have a wide appeal to the vast numbers of teachers and undergraduate and graduate students involved in solid state and quantum physics.

Salford   A. D. BOARDMAN

Physics Programs
Edited by A. D. Boardman
© 1980 John Wiley & Sons Ltd.

CHAPTER 7

Elastic Waves in Crystalline Solids

B. W. JAMES

1. INTRODUCTION

Several fundamental physical properties are related to the propagation of sound waves in solids and an understanding of these processes has led to the development of a number of devices. For example ultrasonic delay lines are widely used in colour television receivers, and diffraction grating dispersive filters are used in radar systems for pulse compression.

The behaviour of sound waves in gases, liquids, amorphous solids, and crystalline solids has been widely investigated. The study of the propagation of sound waves in crystalline solids is the most complex and most interesting of the forms of sound wave to investigate since account must be taken of the anisotropic elastic properties of crystalline solids (Love,[1] Musgrave,[2] and Pollard[3]). In each direction in a crystalline solid there will be three modes of propagation which will, in general, all have different phase velocities, particle motion directions, and energy flow directions. The largest velocity is associated with a longitudinal or nearly longitudinal wave and the other two velocities with transverse or nearly transverse waves. The particle motion directions of the three waves always form an orthogonal set. Pure longitudinal and transverse modes occur in high symmetry directions and in some accidental directions which may be obtained from the known physical properties.

2. TENSOR FORMULATION

2.1 Hooke's law

In order to discuss the elastic properties of anisotropic materials and hence the propagation of sound waves it is necessary to use a tensor formulation of Hooke's law as the stress and strain at a point are given by two second-rank tensors. Hence if σ_{ij} is the stress tensor and e_{kl} is the strain tensor then Hooke's law is written as

$$\sigma_{ij} = c_{ijkl}e_{kl} \quad (i, j, k, l = 1, 2, 3), \tag{1}$$

where c_{ijkl} is a fourth-rank tensor of 81 elements relating 9 stress components to 9 strain components. (Note that summation is assumed for repeated suffices, see Nye[4].)

2.2 The strain tensor

The particle displacements **U** of the strained material determine the nine elements of the general strain tensor E_{kl} and

$$E_{kl} = \frac{\partial U_k}{\partial x_l}, \tag{2}$$

where $\mathbf{U} = \mathbf{i}_1 U_1 + \mathbf{i}_2 U_2 + \mathbf{i}_3 U_3 = \mathbf{i}_i U_i$ and $\mathbf{i}_1, \mathbf{i}_2,$ and $\mathbf{i}_3$ are unit vectors along the Cartesian axes $x_1, x_2,$ and x_3 respectively.

The general tensor E_{kl} consists of a symmetrical part e_{kl} and an antisymmetrical part w_{kl}, where

$$e_{kl} = \frac{1}{2}\left(\frac{\partial U_k}{\partial x_l} + \frac{\partial U_l}{\partial x_k}\right), \tag{3}$$

and

$$w_{kl} = -\frac{1}{2}\left(\frac{\partial U_k}{\partial x_l} - \frac{\partial U_l}{\partial x_k}\right) = -w_{lk}. \tag{4}$$

Now if we first consider a rotation of the material about the origin of the axes without any deformation of the material, then in this rotation the displacement of any point is perpendicular to its radius vector so that

$$U_i x_i = 0 \quad \text{(scalar product)}, \tag{5}$$

or

$$E_{ij}x_i x_j = 0. \tag{6}$$

Since this is true for all x_i the coefficients on the left-hand side must all be zero. Hence

$$E_{ij} = 0 \quad \text{if } i = j; \quad E_{ij} = -E_{ji} \quad \text{if } i \neq j, \tag{7}$$

which is just the condition for E_{ij} to be antisymmetrical. So that in the special case of a rotation of the material about the origin of the axes without deformation, the general strain tensor E_{kl} becomes antisymmetrical, with the rotation of the material about the origin of the axes given by the antisymmetrical part of the tensor, w_{kl}. Thus, in the second case of a combined deformation of the material and of a rotation of it about the origin of the axes, the general strain tensor E_{kl} has a symmetrical part, e_{kl}, which gives the deformation of the material and an antisymmetrical part, w_{kl}, which gives the rotation of the material about the axes. The symmetrical part of the strain tensor (equation (3)) that does not contain rotation about the origin of the axes, and has only six independent elements, is used in the generalization of Hooke's law. This law relates the state of stress of the material to its state of deformation irrespective of the orientation of the material to the axes. Note that the strains $\partial U_i / \partial x_j$, $i = j$, correspond to longitudinal strains and those of the form $\partial U_i / \partial x_j$, $i \neq j$, correspond to shear strains.

2.3 The stress tensor

The stress tensor at a point can also be reduced to six independent components, and is defined in terms of the force per unit area acting across the faces of a unit cube as the volume of the cube tends to zero. The stresses σ_{ii} arise from forces acting normally to the faces and the stresses σ_{ij}, $i \neq j$, are shear stresses in which the force acts parallel to the plane of the face in the direction of the first subscript and across a plane perpendicular to the direction given by the second subscript. Clearly the forces acting across opposite pairs of faces of the cube of material must be equal and opposite in the case of equilibrium and this also applies in the limit under dynamic conditions. Furthermore, in equilibrium there is no rotation of the material, so by considering moments about each axis in turn,

$$\sigma_{ij} = \sigma_{ji}, \tag{8}$$

and again, this also applies in the limit under dynamic conditions. Thus there are just six independent components of stress.

2.4 The matrix notation

Returning to Hooke's law (equation (1)) it is apparent that in place of a possible 81 components of c_{ijkl} the above symmetry of σ_{ij} and e_{kl} lead to at most 36 independent elastic constants. At this point it is appropriate to introduce the more compact two-suffix notation for the elastic constants, that is used in the literature, in which the tensor for the $c_{ijkl}(i, j, k, l = 1, 2, 3)$ is replaced by the matrix c_{ij}, $i, j = 1, 2, \ldots 6$ according to the following scheme of subscript equality:

Tensor notation	11	22	33	23,32	31,13	12,21
Matrix notation	1	2	3	4	5	6.

For example c_{1112} would be replaced by c_{16}. The stress σ_{ij} and the strain e_{kl} are also converted to the matrix notation according to this scheme, except that in the case of the strains a factor of $\frac{1}{2}$ is introduced in the matrix notation when $k \neq l$, so that

$$\sigma_i = c_{ij}e_j, \tag{9}$$

is the matrix form of Hooke's law.

3. ENERGY OF DEFORMATION

The potential energy Φ of the crystal may be expressed as a Taylor series in terms of the strain, and hence

$$\Delta\Phi = \Phi - \Phi(0) = \left(\frac{\partial\Phi}{\partial e_{ij}}\right)_0 e_{ij} + \frac{1}{2}\left(\frac{\partial^2\Phi}{\partial e_{ij}\partial e_{kl}}\right)_0 e_{ij}e_{kl}$$
$$+ \frac{1}{6}\left(\frac{\partial^3\Phi}{\partial e_{ij}\partial e_{kl}\partial e_{mn}}\right)_0 e_{ij}e_{kl}e_{mn} + \ldots, \tag{10}$$

where $\Phi(0)$ is the potential energy of the crystal in the equilibrium unstrained state and $\Delta\Phi$ is the change in potential energy from the equilibrium state to the strained state. Since the deformation is about the equilibrium position, $(\partial\Phi/\partial e_{ij})_0$, and, for small strains, terms in $e_{ij}e_{kl}e_{mn}$ and higher order terms may be neglected. For deformation about the equilibrium position $\Delta\Phi$ must be positive, which means that the right-hand side of equation (10) must be a positive quadratic form with the consequence that

$$\left(\frac{\partial^2\Phi}{\partial e_{ij}\partial e_{kl}}\right)_0 = \left(\frac{\partial^2\Phi}{\partial e_{kl}\partial e_{ij}}\right)_0. \tag{11}$$

Now during the application of the stress σ_{ij} a certain amount of work ΔW will be done to produce the strain e_{ij} and

$$\Delta W = \tfrac{1}{2}\sigma_{ij}e_{ij}. \tag{12}$$

The work ΔW may be equated to the strain energy $\Delta\Phi$ so that

$$\sigma_{ij} = \left(\frac{\partial^2\Phi}{\partial e_{ij}\partial e_{kl}}\right)_0 e_{kl}, \tag{13}$$

and hence

$$c_{ijkl} = \left(\frac{\partial^2\Phi}{\partial e_{ij}\partial e_{kl}}\right)_0 = c_{klij}. \tag{14}$$

which establishes another 15 equalities in the elastic constants and reduces the maximum number of independent constants to 21. It may be noted that the elastic constant tensor c_{ijkl} is equated to the second-order differential of the crystal potential energy function and for this reason they are sometimes referred to as second-order elastic constants. The next term in the Taylor series expansion gives rise to the third-order elastic constants, which give a measure of the anharmonic form of the interatomic forces or their deviation from the harmonic form of an ideal Hooke's law solid.

The number of independent elastic constants for each crystal class and their suffices are given by Nye[4] in graphical form and this is summarized in Table 1.

Table 1. The non-zero elastic constants for the various crystal systems and point groups. (Remember that $c_{ij} = c_{ji}$).

System	Number of point groups	Point group (Schoenflies)	Number of c_{ij}	c_{ij}
Triclinic	2	C_i and C_1	21	
Monoclinic	3	C_{2h}, C_s and C_2	13	$c_{11}, c_{12}, c_{13}, c_{15}, c_{22}, c_{23},$ $c_{25}, c_{33}, c_{35}, c_{44}, c_{46}, c_{55}$ and c_{66}
Orthorhombic	3	D_{2h}, C_{2v} and D_2	9	$c_{11}, c_{12}, c_{13}, c_{22}, c_{23}, c_{33},$ c_{44}, c_{55} and c_{66}
Tetragonal	3	C_{4h}, S_4 and C_4	7	$c_{11} = c_{22}, c_{12}, c_{13} = c_{23},$ $c_{16} = -c_{26}, c_{33}, c_{44} = c_{55}$ and c_{66}
Tetragonal	4	$D_{4h}, D_{2d},$ C_{4v} and D_4	6	$c_{11} = c_{22}, c_{12}, c_{13} = c_{23}, c_{33},$ $c_{44} = c_{55}$ and c_{66}
Trigonal	2	C_{3i} and C_3	7	$c_{11} = c_{22}, c_{12}, c_{13} = c_{23},$ $c_{14} = -c_{24}, -c_{15} = c_{25}, c_{33}$ and $c_{44} = c_{55}$ whilst $c_{46} = 2c_{25}, c_{56} = 2c_{14}$ and $c_{66} = \frac{1}{2}(c_{11} - c_{12})$
Trigonal	3	D_{3d}, C_{3v} and D_3	6	$c_{11} = c_{22}, c_{12}, c_{13} = c_{23},$ $c_{14} = -c_{24}, c_{33}$ and $c_{44} = c_{55}$ whilst $c_{56} = 2c_{14}$ and $c_{66} = \frac{1}{2}(c_{11} - c_{12})$
Hexagonal	7	$D_{6h}, D_{3h}, C_{6v},$ D_6, C_{6h}, C_{3h} and C_6	5	$c_{11} = c_{22}, c_{12}, c_{13} = c_{23}, c_{33}$ and $c_{44} = c_{55}$ whilst $c_{66} = \frac{1}{2}(c_{11} - c_{12})$
Cubic	5	$O_h, T_d, O,$ $T_h,$ and T	3	$c_{11} = c_{22} = c_{33}, c_{12} = c_{13} = c_{23}$ and $c_{44} = c_{55} = c_{66}$

4. THE WAVE EQUATION

The initial step in the derivation of the wave equation and hence of an expression for the velocity of the waves involves, as usual, the equations of motion for an element of the material, but now these must be expressed in the tensor format so that

$$\frac{\partial \sigma_{ij}}{\partial x_i} = \rho u_j, \tag{15}$$

where ρ is the density of the material. Now substituting equations (1) and (3) into (15) eliminates the stress components so that

$$\frac{1}{2} c_{ijkl} \frac{\partial}{\partial x_i} \left(\frac{\partial U_k}{\partial x_l} + \frac{\partial U_l}{\partial x_k} \right) = \rho u_j. \tag{16}$$

If it is now assumed that for an infinite anisotropic elastic solid the solution of equation (16) is a plane wave with constant amplitude **A**, propagation wave number k, with wavefronts normal to a vector with direction cosines l_1, l_2, and l_3, then:

where v is the phase velocity ω/k in the direction given by l_i. The components of the amplitude may be denoted by α_i so that:

$$\mathbf{A} = \mathbf{i}_i \alpha_i. \tag{18}$$

Then when the assumed solution is substituted into equation (16) the following homogeneous equations result:

$$(\Gamma_{jk} - \delta_{jk} \rho v^2) \alpha_j = 0, \tag{19}$$

where

$$\Gamma_{jk} = \tfrac{1}{2} l_i l_l (c_{ijkl} + c_{ijlk}), \tag{20}$$

and

$$\delta_{jk} = 0, \quad \text{for } k \neq j; \qquad \delta_{jk} = 1 \quad \text{for } k = j. \tag{21}$$

These equations have a non-trivial solution only if the secular equation:

$$|\Gamma_{ij} - \delta_{ij} \rho v^2| = 0, \tag{22}$$

is satisfied. This equation is cubic in ρv^2 with three real roots, and from them the velocities v_1, v_2, and v_3 are obtained. The largest velocity corresponds to a longitudinal, or nearly longitudinal, wave and the other two velocities are to transverse or nearly transverse waves. The three values of ρv^2 are, of course, the eigenvalues of the matrix Γ_{jk}, and the corresponding solutions for the displacement vector **A** are the eigenvectors. For each value of v there is a corresponding solution, for which the relative values of the displacement components can be determined from equation (19).

The three displacement vectors or particle motion vectors are:

$$\mathbf{A}_1 = d_1 \mathbf{i}_i a_i; \qquad \mathbf{A}_2 = d_2 \mathbf{i}_i b_i; \qquad \mathbf{A}_3 = d_3 \mathbf{i}_i c_i, \qquad (23)$$

corresponding to three roots ρv_1^2, ρv_2^2, and ρv_3^2 respectively; d_1, d_2, and d_3 are constants dependent on the excitation, and

$$\frac{a_2}{a_1} = \frac{\Gamma_{22}(\Gamma_{11} - \rho v_1^2) - \Gamma_{12}\Gamma_{13}}{\Gamma_{13}(\Gamma_{22} - \rho v_1^2) - \Gamma_{12}\Gamma_{23}}$$

$$\frac{a_3}{a_1} = \frac{\Gamma_{23}(\Gamma_{11} - \rho v_1^2) - \Gamma_{12}\Gamma_{13}}{\Gamma_{12}(\Gamma_{33} - \rho v_1^2) - \Gamma_{23}\Gamma_{13}}, \qquad (24)$$

and

$$a_1^2 + a_2^2 + a_3^2 = 1.$$

There are similar expressions associated with v_2 and v_3 for the b_i and the c_i. The energy flow vector, as defined by Love,[1] is

$$E_i = \sigma_{ij} U_j, \qquad (25)$$

It represents the energy flow per unit time across a surface of unit area normal to this vector. An analogue of the Poynting's vector for the direction of energy flow is obtained when equations (1), (3), and (23) are substituted into (25) and the result is averaged over a cycle, namely:

$$S_i = \frac{1}{2} d_1^2 \frac{\omega^2}{v_1} c_{ijkl} a_j (l_l a_k + l_k a_l), \qquad (26)$$

with similar expressions for the other two waves.

5. THE COMPUTER PROGRAM

5.1 Program description

The program given here permits the calculation of the velocity, the particle motion direction and the energy flow direction of each mode of propagation for any crystal class in any direction, from the appropriate set of elastic constants and the density of the material. This is done by solving equations (22), (24), and (26). To specify a direction in the crystal two angular coordinates are used rather than the direction cosines used in the calculations, as the former are more easily visualized. The angles used are the zenith angle θ and the azimuth angle ϕ, where θ is the angle between the direction and the positive z-axis and ϕ is the angle between the projection of the direction on to the x–y plane and the positive x-axis. The calculations are performed by the program in a region specified by lower and upper angular limits and divided into a number of steps or intervals.

The master segment controls the stepping of the directions within the range given and the subroutine VELOCT is used to calculate the velocities from the elastic constants and density for the direction specified. Within the subroutine VELOCT, after setting up the appropriate equations, two library subroutines FO1AJF and FO2AMF are used to obtain the roots of equation (22), namely the eigenvalues. Both library subroutines are from the NAG library, but any equivalent subroutines could be used. The subroutine FO1AJF gives the Householder reduction of a real symmetric matrix to tridiagonal form for use by FO2AMF. The subroutine FO2AMF calculates all the eigenvalues and eigenvectors of the real symmetric tridiagonal matrix that has been produced by FO1AJF. The velocities v_i are then calculated from the square root of the eigenvalue divided by the density, after which control returns to the master segment. The particle motion direction is obtained from the eigenvectors produced by FO2AMF in VELOCT, instead of from equation (24), and the energy flow direction is obtained by evaluation of equation (26). The subroutine DIRECT is used to obtain the spherical polar directions from the direction cosine data.

The output from the program consists of the input data and, for each direction in the region specified, the three velocities and the associated particle motion directions and energy flow directions. In the output all directions are given both in terms of angles θ and ϕ, and in terms of the direction cosines l_1, l_2, and l_3. If the elastic constants are given in $GN\,m^{-2}$ and the density in $1000\,kg\,m^{-3}$, then the velocity values will appear in $km\,s^{-1}$ since $v = (F/\rho)^{\frac{1}{2}}$ where v is the velocity, F is the effective elastic constant for the mode, and ρ is the density. The effective elastic constant F has the same units as the elastic constants c_{ij} and is a function of them dependent on the mode considered.

5.2 Running the program

The elastic constants and densities of a number of materials are given in Table 2. Following the listing of the program a sample set of data based upon Table 2 is given, together with the associated output produced by a run of the program. The crystallographic axes for a tetragonal material, such as $CaMoO_4$, are an orthogonal set, with the z-axis parallel to the axis of fourfold rotational symmetry. The x- and y-axes are equivalent and they are chosen to be in the direction of the two equal dimensions of the smallest unit cell. A number of general observations may be made for sound wave propagation in tetragonal materials. For propagation along the z-axis $l_1 = 0$, $l_2 = 0$, and $l_3 = 1$, and after appropriate substitution into equations (19) and (22) it is clear that the three waves are all pure modes since the matrix is diagonal, and simple expressions can be obtained for the velocities. For propagation in the x–y plane one of the transverse waves is a pure mode

Table 2. The elastic constants and densities of some crystalline solids. All values of elastic constants are given, in $GN\,m^{-2}$, and only zero values have been omitted from the table. Note that the last row gives the source of reference

Material Point group	CaWO$_4$ C$_{4h}$	CaMoO$_4$ C$_{4h}$	SrMoO$_4$ C$_{4h}$	MgF$_2$ D$_{4h}$	CaSO$_4$, 2H$_2$O C$_{2h}$	NaBrO$_3$ T	Cu O$_h$
c_{11}	143.87	143.92	115.46	124.0	78.6	55.60	169.0
c_{12}	63.501	68.61	59.85	73.0	41.0	16.76	122.0
c_{13}	56.170	48.43	44.36	54.0	26.8	16.76	122.0
c_{14}							
c_{15}					-7.0		
c_{16}	16.355	12.72	12.09				
c_{22}	143.87	143.92	115.46	124.0	62.7	55.60	169.0
c_{23}	56.170	48.43	44.36	54.0		16.76	122.0
c_{24}					24.2		
c_{25}					3.1		
c_{26}	-16.355	-12.72	-12.09				
c_{33}	130.18	125.86	104.19	177.0	72.6	55.60	169.0
c_{34}							
c_{35}					-17.4		
c_{36}							
c_{44}	33.609	36.91	34.99	55.4	9.10	15.08	75.5
c_{45}							
c_{46}					-1.6		
c_{55}	33.609	36.91	34.99	55.4	26.4	15.08	75.5
c_{56}							
c_{66}	45.073	46.07	47.55	97.8	10.44	15.08	75.5
Density kg m^{-3}	6120	4255	4540	3177	2280	3339	8930
Reference	5	6	7	8	9	10	11

with particle motion parallel to the z-axis. According to Neighbours and Schacher,[12] the other two modes of propagation in this plane are both semi-pure. That is to say the direction of energy flow lies within the x–y plane, although it is not parallel to the direction of propagation. They also show that there are two pure mode directions in the x–y plane at 45° to each other, and the amount by which these are rotated from the x-axis depends on the magnitude of c_{16}. In planes defined by these directions and the z-axis, propagation is again by semi-pure modes, and accidental pure modes may occur in either of these planes at certain angles θ from the z-axis. Because of the crystal symmetry these planes are repeated at 90° intervals. By use of the computer program it is possible to investigate these and other features of wave propagation in crystalline materials.

Neglect of the features of energy flow in tetragonal materials can lead to incorrect identification of the wave modes in experimental measurements and consequent incorrect values of elastic constants, as the direction of propagation and the direction of energy flow may differ by up to 60°,

James.[7] These problems can be particularly acute when measurements are carried out on small specimens and unintended reflections can occur which give rise to misleading values for velocity. An extensive discussion of the features of the propagation of sound in solids is given by Musgrave.[2]

6. FURTHER DEVELOPMENT

Apart from the intrinsic interest in the results of this program there are other uses for the VELOCT subroutine of which two examples are given below:

(1) The Debye characteristic temperature θ_0 of a crystalline solid at 0 K may be calculated from the elastic constants and the density at 0 K using the expression:

$$\theta_0 = \frac{h}{k}\left(\frac{9N}{4\pi V}\right)^{\frac{1}{3}}\left(\int_0^{4\pi}\sum_{j=1}^{3}\frac{1}{v_j^3}\frac{d\Omega}{4\pi}\right)^{-\frac{1}{3}}, \qquad (27)$$

where N is Avogadro's number, h is Planck's constant, k is Boltzmann's constant, V is the molar volume at 0 K, and v_j are the velocities of the two transverse and longitudinal waves. The integration is carried out numerically with the aid of the VELOCT subroutine from this program. A program to evaluate θ_0 has been written by Gluyas, Hughes, and James[5] and from their elastic constant data for $CaWO_4$ extrapolated to 0 K they found $\theta_0 = 246.5$ K. In this case, in place of Avogradro's number N, $2N$ is used since $CaWO_4$ is taken to be 'diatomic' and consists of Ca^{2+} and WO_4^{2-} ions.

(2) Another application of the VELOCT subroutine of this program has been in the evaluation of the optimum values of the elastic constants of $CaMoO_4$ from experimental measurements of the velocities of ultrasonic waves (James[6]). The optimum set of elastic constants has been obtained from an over-determined set of measurements by use of the subroutine VELOCT with an iterative minimization subroutine (EO4CCF from the NAG library). The minimization routine was used to adjust the calculated c_{ij} to obtain the lowest value of

$$\text{SUMSQ} = \sum_{i=1}^{n}\left|\frac{v_i^2(\text{calc.})}{v_i^2(\text{meas.})} - 1\right|, \qquad (28)$$

where v_i (calc.) is the value of the velocity calculated by the subroutine VELOCT from the current values of the c_{ij} and the density, v_i (meas.) is the measured value of that velocity, and n is the number of measurements that have been made. It can be shown that, for an appropriate set of velocity measurements taken from suitably orientated specimens, only as many velocities as the number of independent elastic constants have to be measured. However, once a particular orientation of specimen has been produced it is relatively straightforward to measure the velocity of all three modes and then an over-determined set of measurements becomes available for which this method is ideal.

REFERENCES

1. A. E. R. Love, *The Mathematical Theory of Elasticity* (Cambridge University Press, London 1927).
2. M. J. P. Musgrave, *Crystal Acoustics* (Holden Day, London, 1970).
3. H. F. Pollard, *Sound Waves in Solids* (Pion, London, 1977).
4. J. F. Nye, *Physical Properties of Crystals* (Oxford, London, 1957).
5. M. Gluyas, F. D. Hughes, and B. W. James, *J. Phys. D. Appl. Phys.*, **6,** 2025 (1973).
6. B. W. James, *J. Appl. Phys.*, **45,** 3201 (1974).
7. B. W. James, *Phys. Stat. Sol.* (a), **13,** 89 (1972).
8. H. R. Cutler, J. J. Gibson and K. A. McCarthy, *Solid State Commun.*, **6,** 431 (1968).
9. S. Haussuhl, *Z. Krist.*, **122,** 311 (1965).
10. M. Gluyas, R. Hunter and B. W. James, *J. Phys. D. Appl. Phys.*, **8,** 1 (1975).
11. Y. Hiki and A. V. Granato, *Phys. Rev.*, **144,** 411 (1966).
12. J. R. Neighbours and G. E. Schacher, *J. Appl. Phys.*, **38,** 5366 (1967).

VELOCITY PROGRAM

```
C     THE VELOCITY 0F ULTRASONIC WAVES IN ANY DIRECTION IN A SOLID
C     ARE CALCULATED. THE DIRECTIONS IN WHICH THE VELOCITIES ARE
C     CALCULATED ARE SPECIFIED BY THE POLAR DIRECTIONS THETA AND PHI.
C     THE CALCULATIONS ARE DONE FOR A SERIES OF THETA AND PHI VALUES
C     GIVEN BY LOWER ANGULAR LIMITS TL AND PL, BY UPPER ANGULAR LIMITS
C     TU AND PU, AND BY THE NUMBER OF INTERMEDIATE DIRECTIONS NT AND NP.
      REAL LL
      DIMENSION C(6,6),LL(3),V(3),A(3,3),D(3),G(3,3),IA(20),
     1NOT(3,3),SS(3,3)
      DATA NOT(1,1),NOT(1,2),NOT(1,3),NOT(2,1),NOT(2,2),
     1NOT(2,3),NOT(3,1),NOT(3,2),NOT(3,3)/1,6,5,6,2,4,5,4,3/
      PI=3.141592654
      Z=PI/180.0
C     READ INPUT DATA
   23 READ(5,104)IC
      IF (IC-1) 1000,33,1
C     IC IS USED AS A CONTROL INTEGER AS FOLLOWS.
C     FOR IC=2 NEW TITLE,RHO,CIJ,TL,TH,NT,PL,PH,AND NP ARE SOUGHT,
C     IC MUST BE SET EQUAL TO 2 AT THE BEGINNING OF THE DATA.
C     FOR IC=1 NEW TL,TH,NT,PL,PH,AND NP ARE SOUGHT FOR USE WITH
C     THE PREVIOUS RHO AND CIJ.
C     FOR IC=0 RUN OF PROGRAM IS ENDED.
    1 READ(5,100)IA
      READ(5,101)RHO
      WRITE(6,200)IA
      WRITE(6,201)RHO
      WRITE(6,208)
      DO 2 I=1,6
      READ(5,103)(C(I,J),J=1,6)
    2 WRITE(6,202)(C(I,J),J=1,6)
   33 READ(5,102)TL,TH,NT
      READ(5,102)PL,PH,NP
      WRITE(6,203)TL,TH,NT
      WRITE(6,204)PL,PH,NP
      WRITE(6,206)
  100 FORMAT(20A4)
  101 FORMAT(1F0.0)
  102 FORMAT(2F0.0,I0)
  103 FORMAT(6F0.0)
  104 FORMAT(1I0)
  200 FORMAT(1H1,20A4)
  201 FORMAT(1H ,/,  DENSITY = ',F10.4,1H ,'*1000 KGM-3',/)
  202 FORMAT(1H ,'CIJ MATRIX',6F10.4)
  203 FORMAT(1H ,/,' THETA RANGE',F8.4,1H ,'TO',F8.4,1H ,'IN',I6,1H ,
     1'STEPS')
  204 FORMAT(1H ,'PHI RANGE   ',F8.4,1H ,'TO',F8.4,1H ,'IN',I6,1H ,
     1'STEPS')
  205 FORMAT(1H ,/,' MODE VELOCITY',F8.3,1H ,'KM/S')
  206 FORMAT(1H ,///,36X,'THETA        PHI      L1        L2       L3',/)
  207 FORMAT(1H ,/,' ONE EIGENVALUE NEEDS MORE THAN 30 ITERATIONS',/)
  208 FORMAT(1H ,'ELASTIC CONSTANTS GNM-2'
      FNT=FLOAT(NT)
      FNP=FLOAT(NP)
      TS=(TH-TL)/FNT*Z
      PS=(PH-PL)/FNP*Z
      T=TL*Z
```

VELOCITY PROGRAM

```
      DO 3 I=1,NT
      P=PL*Z
      DO 4 J=1,NP
C     CACULATE VELOCITIES V,AND DISPLACEMENT VECTORS A
      IFAIL=1
      CALL VELOCT(RHO,T,P,C,V,LL,A,NOT,IFAIL)
      THETA=T/Z
      PHI=P/Z
      WRITE(6,209)THETA,PHI,LL(1),LL(2),LL(3)
      IF (IFAIL) 6,6,5
    5 WRITE(6,207)
      GOTO 4
    6 CONTINUE
  209 FORMAT(1H ,//,' PROPAGATION DIRECTION ,9X,5F10.4)
      DO 10 NR=1,3
C     CALCULATE ENERGY VECTORS S
      DO 11 IT=1,3
      SUM=0.0
      DO 12 JT=1,3
      DO 13 KT=1,3
      DO 14 LT=1,3
      M=NOT(IT,JT)
      N=NOT(KT,LT)
      SUM=SUM+C(M,N)*A(JT,NR)*(LL(LT)*A(KT,NR)+LL(KT)*A(LT,NR))
   14 CONTINUE
   13 CONTINUE
   12 CONTINUE
      SS(IT,NR)=SUM
   11 CONTINUE
   10 CONTINUE
C     PRINT FOR EACH VELOCITY THE PARTICLE MOTION AND
C     ENERGY FLOW DIRECTIONS AND FINALLY THE VELOCITY
      DO 15 NR=1,3
      WRITE(6,205)V(NR)
      CALL DIRECT(A,NR,210,1E-4)
      CALL DIRECT(SS,NR,211,1E-4)
   15 CONTINUE
    4 P=P+PS
    3 T=T+TS
      GOTO 23
C     CONTROL NOW RETURNS TO THE BEGINNING OF THE PROGRAM.
C     A NEW VALUE OF THE CONTROL INTEGER IS USED FOLLOWS.
C     TO END THE RUN OF THE PROGRAM, IC=0
C     TO ENABLE NEW RANGES OF THETA AND PHI TO BE SELECTED, IC=1
C     TO RUN THE PROGRAM FOR ANOTHER MATERIAL, IC=2
 1000 CONTINUE
      STOP
      END
      SUBROUTINE VELOCT(RHO,T,P,C,V,LL,A,NOT,IFAIL)
C     RHO IS THE DENSITY,T AND P ARE THE POLAR COORDINATES
C     THETA AND PHI IN RADIANS C IS A 6*6 ARRAY OF THE
C     ELASTIC CONSTANTS,V IS AN ARRAY OF THE 3 VELOCITIES
C     AND LL IS AN ARRAY OF THE 3 DIRECTION COSINES AND A IS
C     A 3*3 ARRAY OF THE EIGENVECTORS
      REAL LL
      DIMENSION NOT(3,3),C(6,6),G(3,3),A(3,3),LL(3),V(3),D(3),E(3)
```

VELOCITY PROGRAM

```
C     CALCULATE DIRECTION COSINES OF PROPAGATION DIRECTION
      LL(1)=SIN(T)*COS(P)
      LL(2)=SIN(T)*SIN(P)
      LL(3)=COS(T)
C     SET UP CHRISTOFFEL MATRIX
      DO 1 JS=1,3,1
      DO 2 KS=1,3,1
      SUMS=0.0
      DO 3 IS=1,3,1
      DO 4 LS=1,3,1
      MS=NOT(IS,JS)
      NS=NOT(KS,LS)
      SUMS=SUMS+LL(IS)*LL(LS)*C(MS,NS)
    4 CONTINUE
    3 CONTINUE
      G(JS,KS)=SUMS
      IF(JS.NE.KS) G(KS,JS)=G(JS,KS)
    2 CONTINUE
    1 CONTINUE
      CALL F01AJF(3,2.0**(-218),G,3,D,E,A,3)
      CALL F02AMF(3,2.0**(-37),D,E,A,3,IFAIL)
      IF (IFAIL) 5,5,6
    5 DO 7 JS=1,3
    7 V(JS)=SQRT(D(JS)/RHO)
    6 RETURN
      END
      SUBPOUTINE DIRECT(A,NR,JW,EPS)
C     FOR A(I,R) WHERE I=1,2,3 IN WHICH THE A(I,R) ARE NOT
C     NECESSARILY NORMALIZED THE CORRESPONDING SPHEPICAL
C     POLAR CO-ORDINATES THETA AND PHI ARE CALCULATED.THETA
C     PHI AND THE NORMALIZED DIRECTION COSINES APE PRINTED
C     JW CONTROLS THE WRITE STATEMENT THAT IS USED
      REAL L
      DIMENSION L(3),A(3,3)
      PI=3.141592654
      Z=PI/180.0
      S=0.0
      DO 1 ID=1,3
    1 IF(ABS(A(ID,NR)).GT.S) S=ABS(A(ID,NR))
C     REDUCE RANGE OF DIRECTION COSINES TO 0 TO 1
      DO 2 ID=1,3
    2 L(ID)=A(ID,NR)/S
      S=0.0
      DO 3 ID=1,3
    3 S=S+L(ID)**2
      S=SQRT(S)
C     NORMALIZE DIRECTION COSINES
      DO 4 ID=1,3
    4 L(ID)=L(ID)/S
C     CALCULATE THETA AND PHI FROM NORMALIZED DIRECTION COSINES
      IF(ABS(L(3)).GT.EPS) GO TO 5
      THETA=PI/2.0
      GO TO 7
    5 IF(ABS(L(3)).LT.(1.0-EPS)) GO TO 6
      THETA=0.0
      PHI=0.0
```

VELOCITY PROGRAM

```
C     PHI HAS NO MEANING FOR THETA=0.0 THEREFORE SET PHI=0.0
      GO TO 15
    6 THETA=ATAN(SQRT(1.0-L(3)**2)/L(3))
C     NOTE THETA DEFINED 0.0 TO PI WHEREAS ATAN RANGE -PI/2 TO +PI/2
      IF (THETA.GT.0.0) GO TO 7
      THETA=PI+THETA
    7 PHI=0.0
      IF (L(1)) 13,8,9
    8 IF (L(2)) 10,15,11
    9 IF (L(2)) 12,14,14
   10 PHI=PI
   11 PHI=PHI+PI/2.0
      GO TO 15
   12 PHI=PI
   13 PHI=PHI+PI
   14 PHI=PHI+ATAN(L(2)/L(1))
   15 CONTINUE
      THETA=THETA/2
      PHI=PHI/2
      IF (JW-210) 16,16,17
   16 WRITE(6,210)THETA,PHI,L(1),L(2),L(3)
      GOTO 18
   17 WRITE(6,211)THETA,PHI,L(1),L(2),L(3)
  210 FORMAT(1H ,'PARTICLE MOTION',15X,5F10.4)
  211 FORMAT(1H ,'ENERGY FLOW',19X,5F10.4)
   18 CONTINUE
      RETURN
      END
```

DATA FOR THE VELOCITY PROGRAM

```
2
CALCIUM MOLYBDATE DATA FROM B W JAMES J APPL PHYS VOL 45 3201 (1974)
4.255
143.92  68.61  48.43  0.0  0.0  12.72
68.61  143.92  48.43  0.0  0.0  -12.72
48.43  48.43  125.86  0.0  0.0  0.0
0.0  0.0  0.0  39.91  0.0  0.0
0.0  0.0  0.0  0.0  39.91  0.0
12.72  -12.72  0.0  0.0  0.0  46.07
90.0  95.0  1
0.0  90.0  6
0
```

RESULTS FROM THE VELOCITY PROGRAM

CALCIUM MOLYBDATE DATA FROM B W JAMES J APPL PHYS VOL 45 3201 (1974)

DENSITY =   4.2550 *1000 KGM-3

ELASTIC CONSTANTS GNM-2

CIJ MATRIX	143.9200	68.6100	48.4300	0.0000	0.0000	12.7200
CIJ MATRIX	68.6100	143.9200	48.4300	0.0000	0.0000	-12.7200
CIJ MATRIX	48.4300	48.4300	125.8600	0.0000	0.0000	0.0000
CIJ MATRIX	0.0000	0.0000	0.0000	39.9100	0.0000	0.0000
CIJ MATRIX	0.0000	0.0000	0.0000	0.0000	39.9100	0.0000
CIJ MATRIX	12.7200	-12.7200	0.0000	0.0000	0.0000	46.0700

THETA RANGE 90.0000 TO 95.0000 IN   1 STEPS
PHI RANGE   0.0000 TO 90.0000 IN   6 STEPS

	THETA	PHI	L1	L2	L3
PROPAGATION DIRECTION	90.0000	0.0000	1.0000	0.0000	-0.0000
MODE VELOCITY 3.063 KM/S					
PARTICLE MOTION	0.0000	0.0000	0.0000	-0.0000	1.0000
ENERGY FLOW	90.0000	360.0000	1.0000	-0.0000	-0.0000
MODE VELOCITY 3.232 KM/S					
PARTICLE MOTION	90.0000	97.2868	-0.1268	0.9919	0.0000
ENERGY FLOW	90.0000	328.9673	0.8569	-0.5155	-0.0000
MODE VELOCITY 5.849 KM/S					
PARTICLE MOTION	90.0000	7.2868	0.9919	0.1268	-0.0000
ENERGY FLOW	90.0000	10.4099	0.9835	0.1807	-0.0000
PROPAGATION DIRECTION	90.0000	15.0000	0.9659	0.2588	-0.0000
MODE VELOCITY 2.749 KM/S					
PARTICLE MOTION	90.0000	289.5796	0.3351	-0.9422	0.0000
ENERGY FLOW	90.0000	347.2489	0.9753	-0.2207	-0.0000
MODE VELOCITY 3.063 KM/S					
PARTICLE MOTION	0.0000	0.0000	0.0000	0.0000	1.0000
ENERGY FLOW	90.0000	15.0000	0.9659	0.2588	-0.0000
MODE VELOCITY 6.091 KM/S					
PARTICLE MOTION	90.0000	199.5796	-0.9422	-0.3351	0.0000
ENERGY FLOW	90.0000	21.1168	0.9328	0.3603	-0.0000
PROPAGATION DIRECTION	90.0000	30.0000	0.8660	0.5000	-0.0000
MODE VELOCITY 2.597 KM/S					
PARTICLE MOTION	90.0000	298.8262	0.4822	-0.8761	-0.0000
ENERGY FLOW	90.0000	38.6282	0.7812	0.6243	-0.0000

RESULTS FROM THE VELOCITY PROGRAM

```
MODE VELOCITY    3.063 KM/S
PARTICLE MOTION              0.0000     0.0000     0.0000     0.0000    1.0000
ENERGY FLOW                 90.0000    30.0000     0.8660     0.5000   -0.0000

MODE VELOCITY    6.157 KM/S
PARTICLE MOTION             90.0000   208.8262    -0.8761    -0.4822    0.0000
ENERGY FLOW                 90.0000    28.4542     0.8792     0.4765   -0.0000


PROPAGATION DIRECTION       90.0000    45.0000     0.7071     0.7071   -0.0000

MODE VELOCITY    2.919 KM/S
PARTICLE MOTION             90.0000   308.7462     0.6259    -0.7799   -0.0000
ENERGY FLOW                 90.0000    77.4021     0.2181     0.9759   -0.0000

MODE VELOCITY    3.063 KM/S
PARTICLE MOTION              0.0000     0.0000     0.0000    -0.0000    1.0000
ENERGY FLOW                 90.0000    45.0000     0.7071     0.7071   -0.0000

MODE VELOCITY    6.011 KM/S
PARTICLE MOTION             90.0000   218.7462    -0.7799    -0.6259    0.0000
ENERGY FLOW                 90.0000    36.4858     0.8040     0.5946   -0.0000


PROPAGATION DIRECTION       90.0000    60.0000     0.5000     0.8660   -0.0000

MODE VELOCITY    3.063 KM/S
PARTICLE MOTION              0.0000     0.0000    -0.0000     0.0000    1.0000
ENERGY FLOW                 90.0000    60.0000     0.5000     0.8660   -0.0000

MODE VELOCITY    3.416 KM/S
PARTICLE MOTION             90.0000   323.6264     0.8052    -0.5930    0.0000
ENERGY FLOW                 90.0000    85.2760     0.0824     0.9966   -0.0000

MODE VELOCITY    5.743 KM/S
PARTICLE MOTION             90.0000   233.6264    -0.5930    -0.8052    0.0000
ENERGY FLOW                 90.0000    50.5144     0.6359     0.7718   -0.0000


PROPAGATION DIRECTION       90.0000    75.0000     0.2588     0.9659   -0.0000

MODE VELOCITY    3.063 KM/S
PARTICLE MOTION              0.0000     0.0000     0.0000     0.0000    1.0000
ENERGY FLOW                 90.0000    75.0000     0.2588     0.9659   -0.0000

MODE VELOCITY    3.592 KM/S
PARTICLE MOTION             90.0000   346.9417     0.9741    -0.2259   -0.0000
ENERGY FLOW                 90.0000    67.5301     0.3822     0.9241   -0.0000

MODE VELOCITY    5.635 KM/S
PARTICLE MOTION             90.0000   256.9417    -0.2259    -0.9741    0.0000
ENERGY FLOW                 90.0000    78.0494     0.2071     0.9783   -0.0000
```

Physics Programs
Edited by A. D. Boardman
© 1980 John Wiley & Sons Ltd.

CHAPTER 8

A Computer-assisted Tutorial in Time-independent Non-degenerate Perturbation Theory

D. J. MARTIN

1. INTRODUCTION

Probably the most widely used approximate method in quantum mechanics is that known as perturbation theory. The idea of the following exercise is for the student to calculate the energies of a particle in a particular form of one-dimensional potential well, first with the computer program—which uses a numerical iterative technique to give very accurate results—and then by employing first- and second-order perturbation theory. The student has the incentive of trying to achieve agreement with the computer—rather as the theoretician seeks to achieve results in conformity with experiments.

The form of the potential—an infinite potential well with an extra potential (the perturbation) in part of the well (see Figure 1)—is chosen so that the Schrödinger equation can be solved numerically to a high degree of accuracy and that the perturbation theory calculation is relatively straightforward.

2. BASIC QUANTUM MECHANICS

2.1 Introduction

The central theoretical problem in treating any microscopic system—in which quantum-mechanical effects are significant—is that of finding the solution to the appropriate Schrödinger equation. For a single particle of mass M_p in a time-independent scalar field $V(\mathbf{r})$ the (time-independent) Schrödinger equation is

$$\left[-\frac{\hbar^2}{2M_p}\nabla^2 + V(\mathbf{r})\right]\psi(\mathbf{r}) = E\psi(\mathbf{r}), \tag{1}$$

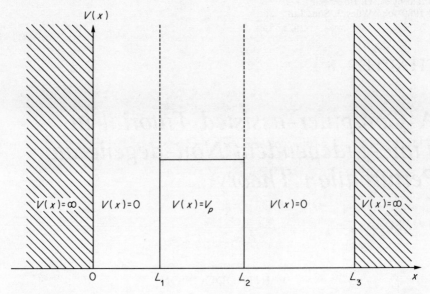

Figure 1. The form of the perturbed infinite potential well. V_p can be >0 or <0

where $\hbar = h/2\pi$, h is Planck's constant ($\hbar = 1.056 \times 10^{-34}$ J s) and ∇^2 is the Laplacian operator (sometimes written Δ) which in Cartesian coordinates is

$$\nabla^2 = \left\{ \frac{\partial^2}{\partial x^2} + \frac{\partial^2}{\partial y^2} + \frac{\partial^2}{\partial z^2} \right\}, \tag{2}$$

$\psi(\mathbf{r})$ is the wave function of the particle. If $\psi(\mathbf{r})$ is known then all possible information about the system can be obtained because the probability density at $\mathbf{r}$ is $\psi^*(\mathbf{r}) \cdot \psi(\mathbf{r})$, where $\psi^*(\mathbf{r})$ is the complex conjugate of $\psi(\mathbf{r})$ · Here, E is the energy of the particle but not all values of E are necessarily possible. In fact it can be shown that if the particle is in a bound state (i.e. $\psi(\mathbf{r}) \to 0$ for $r \to \infty$) then E can only take discrete values, i.e. the energy is quantized. For a given $V(\mathbf{r})$ there are generally a number of different $\psi(\mathbf{r})$ and corresponding energies E which will be distinguished with the subscript n in the form $\psi_n(\mathbf{r})$ and E_n. The expression $[-(\hbar^2/2M_p)\nabla^2 - V(\mathbf{r})]$ is called the Hamiltonian ($\mathscr{H}$) and the time-independent Schrödinger equation can be written as $\mathscr{H}\psi_n(\mathbf{r}) = E_n\psi_n(\mathbf{r})$.

Typically M_p and $V(\mathbf{r})$ are known so that it remains to solve equation (1) for $\psi_n(\mathbf{r})$ and to compare E_n with the results of experiment. There are very few problems for which an exact analytical solution has been found for

$\psi_n(\mathbf{r})$. Indeed, in the majority of cases it is necessary to accept an approximate solution where, of course, approximate should not be confused with inaccurate (i.e. $\sin(x) \approx x$ with high accuracy provided $x \ll 1$.)

Probably the approximate method most widely used in quantum theory is that known as perturbation theory. This method is applicable if the Hamiltonian consists of two parts

$$\mathcal{H} = \mathcal{H}_0 + \mathcal{H}_p, \tag{3}$$

where $\mathcal{H}_p$ has only a small effects and is called the perturbation and where the Schrödinger equation, without $\mathcal{H}_p$, can be solved and all the possible $\psi_n^{(0)}(\mathbf{r})$ and corresponding energies $E_n^{(0)}$ can be found. (The (0) superscript on $\psi_n^{(0)}(\mathbf{r})$ and $E_n^{(0)}$ is used to indicate that they correspond to the unperturbed Hamiltonian $\mathcal{H}_0$.) In the absence of the perturbation $\mathcal{H}_p$ equation (1) is therefore:

$$\mathcal{H}_0 \psi_n^{(0)}(\mathbf{r}) = E_n^{(0)} \psi_n^{(0)}(\mathbf{r}). \tag{4}$$

The simplest form of perturbation theory applies when:

(1) $\mathcal{H}_0$ and $\mathcal{H}_p$ do not vary with time; and
(2) the solutions for the unperturbed system are non-degenerate, i.e. for a given energy $E_n^{(0)}$ there is a unique $\psi_n^{(0)}(\mathbf{r})$.

In this exercise we will confine our attention to quantized energy levels—which simplifies the notation. In fact, once the basic ideas have been grasped, the extension of perturbation theory to more general cases, where degeneracy occurs and/or $\mathcal{H}_p$ varies with time, is relatively simple.

2.2 Time-independent non-degenerate perturbation theory

In order to apply perturbation theory we must know $\psi_n^{(0)}(\mathbf{r})$ and $E_n^{(0)}$—the unperturbed wave functions and energies. Then, even if we cannot solve the Schrödinger equation in the presence of the perturbation $\mathcal{H}_p$, i.e.

$$(\mathcal{H}_0 + \mathcal{H}_p)\psi_n(\mathbf{r}) = E_n \psi_n(\mathbf{r}), \tag{5}$$

perturbation theory enables us to make an accurate estimate of $\psi_n(\mathbf{r})$ and E_n in terms of $\psi_n^{(0)}(\mathbf{r})$ and $E_n^{(0)}$, provided that the effects of $\mathcal{H}_p$ are small.

A full explanation and justification of perturbation theory is beyond the scope of this chapter (the subject is discussed in most intermediate-level textbooks on quantum theory), but the basis of the method is to express both E_n and $\psi_n(\mathbf{r})$—the perturbed energy and the perturbed wave function—as power series in what might be called the 'strength' of $\mathcal{H}_p$ (which we will write as λ_p, a dimensionless number), i.e.

$$E_n = \varepsilon_0 + \lambda_p / \varepsilon_1 + \lambda_p^2 / \varepsilon_2 + \ldots,$$
$$\psi_n(\mathbf{r}) = \chi_0(\mathbf{r}) + \lambda_p / \chi_1(\mathbf{r}) + \lambda_p^2 / \chi_2(\mathbf{r}) + \ldots. \tag{6}$$

Remembering that for $\mathscr{H}_p = 0$ we know the energy ($= E_n^0$) and the wave function ($= \psi_n^{(0)}(\mathbf{r})$) we can rewrite equation (6) as

$$E_n = E_n^{(0)} + \lambda_p/\varepsilon_1 + \lambda_p^2/\varepsilon_2 + \ldots,$$

$$\psi_n(\mathbf{r}) = \psi_n^{(0)}(\mathbf{r}) + \lambda_p/\chi_1(\mathbf{r}) + \lambda_p^2/\chi_2(\mathbf{r}) + \ldots. \tag{7}$$

The second term on the right-hand side of equation (7) is called the first-order correction to the energy or wave function, the third term is called the second-order correction, etc. In principle, perturbation theory enables us to evaluate all the terms in these infinite series and hence to find E_n and $\psi_n(\mathbf{r})$ exactly. However, the complexity of evaluating the terms increases with their order and, in practice, the calculation is limited to the first few terms. Nevertheless, provided λ_p is small, only the first few terms in the series will be significant, and the results for E_n and $\psi_n(\mathbf{r})$ will be accurate.

2.3 The first-order correction to the energy

If $\mathscr{H}_p$ is very small then $\psi_n(\mathbf{r})$ will be almost identical to $\psi_n^{(0)}(\mathbf{r})$. The first-order correction to the energy is found by assuming that any change in the wave function is negligible. The change in the energy will then be

$$\lambda_p\varepsilon_1 = E_n^{(1)} = E_n \approx E_n^{(0)} = \frac{\underset{\text{all space}}{\int} \psi_n^{(0)*}(\mathbf{r})\mathscr{H}_p\psi_n^{(0)}(\mathbf{r})\,d\mathbf{r}}{\underset{\text{all space}}{\int} \psi_n^{(0)*}(\mathbf{r})\psi_n^{(0)}(\mathbf{r})\,d\mathbf{r}} \tag{8}$$

which is the usual expression for the average value of an observable. (If you are unfamiliar with this type of expression then consider the following argument. Suppose $\mathscr{H}_p$ is an extra contribution to the potential, say $V_p(\mathbf{r})$. If the wave function is $\psi_n^{(0)}(\mathbf{r})$ then the probability density at $\mathbf{r}$ is $\psi_n^{(0)*}(\mathbf{r})\psi_n^{(0)}(\mathbf{r})$. The expression given above for the first-order correction to the energy is simply the probability density at $\mathbf{r}$, multiplied by the change in potential energy at $\mathbf{r}$, integrated, i.e. averaged, over all space, and divided by the total probability of the particle being anywhere in space.) It is convenient to choose $\psi_n^{(0)}(\mathbf{r})$ such that

$$\int_{\text{all space}} \psi_n^{(0)*}(\mathbf{r})\psi_n^{(0)}(\mathbf{r})\,d\mathbf{r} = 1 \tag{9}$$

and it will be assumed in the following that this is always the case. (The wave functions are then said to be normalized.) Then, to first order, equation (8) becomes

$$E_n \approx E_n^{(0)} + \int_{\text{all space}} \psi_n^{(0)*}(\mathbf{r})\mathscr{H}_p\psi_n^{(0)}(\mathbf{r})\,d\mathbf{r} \tag{10}$$

and, since $E_n^{(0)}$, $\psi_n^{(0)}(\mathbf{r})$, and $\mathcal{H}_p$ are known, the energy can be found to first order, provided that the integral can be evaluated.

2.4 The first-order correction to the wave function and the second-order correction to the energy

As explained above, the first-order correction to the energy is derived from the uncorrected ('zeroth-order') wave function. The second-order correction to the energy is derived from the first-order corrected wave function, and so on. The first-order correction to the wave function consists of an appropriate linear combination of all the *other* unperturbed wave functions, i.e. to first order:

$$\psi_n(\mathbf{r}) = \psi_n^{(0)}(\mathbf{r}) + \sum_{\substack{m \neq n \\ m=1}}^{\infty} a_m \psi_m^{(0)}(\mathbf{r}). \tag{11}$$

It is possible to show that the coefficients a_m, which determine the amount of wave function $\psi_m^{(0)}(\mathbf{r})$ to be added, are given by

$$a_m = \frac{\displaystyle\int_{\text{all space}} \psi_m^{(0)*}(\mathbf{r}) \mathcal{H}_p \psi_n^{(0)}(\mathbf{r}) \, d\mathbf{r}}{(E_n^{(0)} - E_m^{(0)})}, \tag{12}$$

and the resultant second-order correction to the energy is given by

$$\lambda_p^2 \varepsilon_2 = E_n^{(2)} = E_n - E_n^{(0)} - E_n^{(1)} = \sum_{\substack{m \neq n \\ m=1}}^{\infty} \frac{\left| \displaystyle\int_{\text{all space}} \psi_m^{(0)*}(\mathbf{r}) \mathcal{H}_p \psi_n^{(0)}(\mathbf{r}) \, d\mathbf{r} \right|^2}{(E_n^{(0)} - E_m^{(0)})}. \tag{13}$$

The full expression for the perturbed energy, correct to second order is therefore

$$E_n = E_n^{(0)} + \int_{\text{all space}} \psi_n^{(0)*}(\mathbf{r}) \mathcal{H}_p \psi_n^{(0)}(\mathbf{r}) \, d\mathbf{r}$$

$$+ \sum_{\substack{m \neq n \\ m=1}}^{\infty} \frac{\left| \displaystyle\int_{\text{all space}} \psi_m^{(0)*}(\mathbf{r}) \mathcal{H}_p \psi_n^{(0)}(\mathbf{r}) \, d\mathbf{r} \right|^2}{(E_n^{(0)} - E_m^{(0)})}. \tag{14}$$

Note that the first-order term for the energy only involves a single integral; its evaluation is therefore much simpler than that of the second-order term which is an infinite series of integrals involving all the unperturbed states.

3. A PARTICLE IN A ONE-DIMENSIONAL INFINITE POTENTIAL WELL

3.1 Unperturbed potential well

In one dimension the Schrödinger equation for a particle of mass M_p in a region $0 < x < L_3$, say, where $V(x) = 0$, is

$$\frac{-\hbar^2}{2M_p}\frac{d^2}{dx^2}\psi_n^{(0)}(x) = E_n^{(0)}\psi_n^{(0)}(x). \tag{15}$$

Because $\hbar$ is very small, quantum effects are usually only important in systems with atomic dimensions and masses. (A mass of $1\,kg$ in an infinite potential well of width $1\,m$ would have a ground state energy of $5.5 \times 10^{-63}\,J$, which corresponds to a r.m.s. velocity of $3.3 \times 10^{-34}\,m\,s^{-1}$; an electron of mass $9.11 \times 10^{-31}\,kg$ in an infinite potential well of width $0.1\,nm$—typical of atomic dimensions—would have a ground state energy of $1.19 \times 10^{-17}\,J = 75\,eV$, which corresponds to a r.m.s. velocity of $5.1 \times 10^6\,m\,s^{-1}$.) Rather than enter numbers such as 9.11×10^{-31} (the mass of an electron in kg) into the computer for M_p we will use the atomic system of units in the following.

If lengths are expressed as multiples of the Bohr radius (1 Bohr radius $= 4\pi\varepsilon_0\hbar^2/m_e e^2 = 5.29 \times 10^{-11}\,m$) and energies and potentials are expressed as multiples of the Rydberg (1 Rydberg $= -$ the ground state energy of the electron in the hydrogen atom $= m_e e^4/32\pi\varepsilon_0^2\hbar^2 = 2.19 \times 10^{-18}\,J = 13.6\,eV$) then equation (1), in dimensionless form, is

$$\left[-\frac{1}{M_p}\nabla^2 + V(\mathbf{r})\right]\psi(\mathbf{r}) = E\psi(\mathbf{r}), \tag{16}$$

where $M_p' = M_p/m_e$. For the one-dimensional infinite potential well (see Figure 2) equation (16) becomes (dropping the prime on M_p')

$$-\frac{1}{M_p}\frac{d^2}{dx^2}\psi_n^{(0)}(x) = E_n^{(0)}\psi_n^{(0)}(x). \tag{17}$$

The general solution to (17) is

$$\psi_n^{(0)}(x) = A\,e^{ikx} + B\,e^{-ikx}, \tag{18}$$

where $k^2 = E_n^{(0)}M_p$.

We must impose the boundary conditions that $\psi_n^{(0)}(x) \to 0$, both for $x \to 0$ and $x \to L_3$, because the particle would require an infinite energy to penetrate the region of infinite potential. Thus $B = -A$ and $\psi_n^{(0)}(x) =$

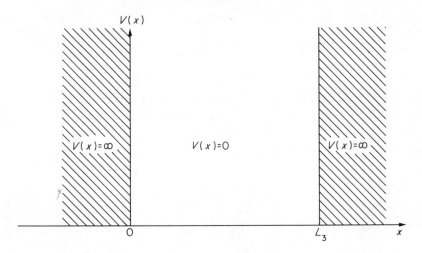

Figure 2. The unperturbed infinite potential well

$C \sin kx$ with $C = 2iA$. But $\psi_n^{(0)}(L_3) = C \sin(kL_3) = 0$, hence

$$k = \frac{n\pi}{L_3}, \qquad n = 1, 2, 3, \ldots,$$

$$E_n^{(0)} = \frac{k^2}{M_p} = \frac{n^2\pi^2}{M_p L_3^2},$$

$$\psi_n^{(0)}(x) = C \sin\left(\frac{n\pi x}{L_3}\right). \tag{19}$$

These are the results for the unperturbed energies and wave functions (see Figure 3). The expressions given earlier for first- and second-order perturbation theory required normalized wave functions so C must be adjusted to ensure this. This is done quite simply by requiring that

$$\int_0^{L_3} C^* \sin\left(\frac{n\pi x}{L_3}\right) C \sin\left(\frac{n\pi x}{L_3}\right) dx = C^2 \left|\frac{x}{2} - \frac{L_3}{4n\pi} \sin\frac{2n\pi x}{L_3}\right|_0^{L_3} = C^2 \frac{L_3}{2} = 1,$$

$$\tag{20}$$

i.e.

$$\psi_n^{(0)}(x) = \sqrt{\frac{2}{L_3}} \sin\left(\frac{n\pi x}{L_3}\right) \qquad n = 1, 2, 3, \ldots \tag{21}$$

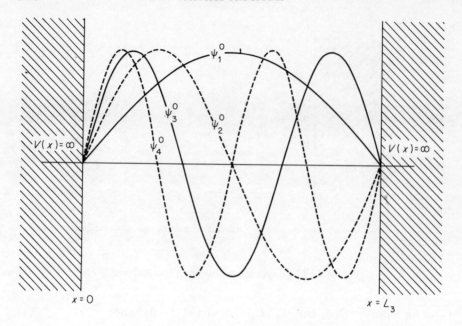

Figure 3. The unperturbed wave functions.

The case $V_p = 0$ should be tried, when running the computer program, detailed in section 4, to check that the computed energies agree with the standard expression

$$E_n^{(0)} = \frac{n^2 \pi^2}{M_p L_3^2}, \qquad n = 1, 2, 3, \ldots \tag{22}$$

for your chosen values of M_p and L_3.

3.2 The first-order correction to the energy for a particle in an infinite one-dimensional potential well with a perturbation V_p

The general expression for the first-order correction to the nth energy level is

$$E_n^{(1)} = \int\limits_{\text{all space}} \psi_n^{(0)*}(\mathbf{r}) \mathcal{H}_p \psi_n^{(0)}(\mathbf{r}) \, d\mathbf{r}. \tag{23}$$

The $\psi_n^{(0)}(\mathbf{r})$ are the normalized unperturbed wave functions found above. The integration is carried out over all space where $\mathcal{H}_p$ is finite—in this case between L_1 and L_2. $\mathcal{H}_p$ itself is the extra term in the Hamiltonian—here simply a constant V_{p}. Equation (23) therefore becomes

$$
\begin{aligned}
E_n^{(1)} &= \int_{L_1}^{L_2} \sqrt{\frac{2}{L_3}} \sin\left(\frac{n\pi x}{L_3}\right) V_{\mathrm{p}} \sqrt{\frac{2}{L_3}} \sin\left(\frac{n\pi x}{L_3}\right) \mathrm{d}x \\
&= \frac{2V_{\mathrm{p}}}{L_3} \int_{L_1}^{L_2} \sin^2\left(\frac{n\pi x}{L_3}\right) \mathrm{d}x \\
&= \frac{V_{\mathrm{p}}}{L_3}(L_2 - L_1) - \frac{V_{\mathrm{p}}}{2n\pi}\left\{\sin\left(\frac{2\pi n L_2}{L_3}\right) - \sin\left(\frac{2\pi n L_1}{L_3}\right)\right\}. \quad (24)
\end{aligned}
$$

As V_{p} increases this energy will increase—as we would expect. However, the rate of increase will be different for different energy levels, because the probability of the particle being between L_1 and L_2 (and therefore 'seeing' V_{p}) will change with n. For example, if L_1 and L_2 are near the centre of the well the perturbation will affect the $n = 1$ level (in which the particle has a high probability of being near the centre) much more than the $n = 2$ level (where the particle has a low probability of being near the centre).

3.3 The second-order correction to the energy for a particle in an infinite one-dimensional potential well with a perturbation V_{p}

The general expression for the second-order correction to the energy of the nth level is

$$
E_n^{(2)} = \sum_{\substack{m \neq n \\ m=1}}^{\infty} \frac{\left| \int_{\text{all space}} \psi_m^{(0)*}(\mathbf{r}) \mathcal{H}_{\mathrm{p}} \psi_n^{(0)}(\mathbf{r}) \, \mathrm{d}\mathbf{r} \right|^2}{(E_n^{(0)} - E_m^{(0)})} = \sum_{\substack{m \neq n \\ m=1}}^{\infty} E_{mn}^{(2)} \quad (25)
$$

In the present case

$$
\begin{aligned}
I_{mn} &= \int_{\text{all space}} \psi_m^{0*}(\mathbf{r}) \mathcal{H}_{\mathrm{p}} \psi_n^{(0)}(\mathbf{r}) \, \mathrm{d}\mathbf{r} \\
&= \int_{L_1}^{L_2} \sqrt{\frac{2}{L_3}} \sin\left(\frac{m\pi x}{L_3}\right) V_{\mathrm{p}} \sqrt{\frac{2}{L_3}} \sin\left(\frac{n\pi x}{L_3}\right) \mathrm{d}x \\
&= \frac{2V_{\mathrm{p}}}{L_3} \int_{L_1}^{L_2} \sin\left(\frac{m\pi x}{L_3}\right)\sin\left(\frac{n\pi x}{L_3}\right) \mathrm{d}x \\
&= \frac{V_{\mathrm{p}}}{L_3}\left\{\frac{\sin(\alpha L_2) - \sin(\alpha L_1)}{\alpha} - \frac{\sin(\beta L_2) - \sin(\beta L_1)}{\beta}\right\}, \quad (26)
\end{aligned}
$$

where $\alpha = \pi(n - n)/L_3$, $\beta = \pi(n + n)/L_3$ and $|m| \neq |n|$.

4. THE COMPUTER PROGRAM

4.1 Basic features

The computer program calculates, to very high accuracy, the four lowest energy states of a particle of mass M_p, in a one-dimensional infinite potential well with an extra constant potential V_p in part of it. The program allows you to choose the particle mass M_p, the width of the finite potential well ($= L_3$), the position where the potential V_p starts ($= L_1$) and ends ($= L_2$), and the potential V_p itself. The basic procedure used to obtain the exact results is outlined below.

For $0 < x < L_1$ and $L_2 < x < L_3$, $V(x) = 0$ and

$$\psi(x) = A \exp(ik_1x) + B \exp(-ik_1x), \qquad k_1 = \sqrt{M_pE}. \tag{27}$$

Similarly, for $L_1 < x < L_2$, $V(x) = V_p$ and

$$\psi(x) = A' \exp(ik_2x) + B' \exp(-ik_2x), \qquad k_2 = \sqrt{M_p(E - V_p)}. \tag{28}$$

Here, $\psi(0) = 0$ is a necessary boundary condition, but this still leaves the assignment of $[d\psi(x)/dx]_{x=0}$. It is chosen to be unity, which is a compromise value resulting in $\psi(x)$ and $d\psi(x)/dx$ being of reasonable size over the whole system. (Note that in this computer method of solution, $\psi(x)$ is not, and need not, be normalized. Also this choice of $d\psi(x)/dx$ is reasonable only provided V_p has not a very extreme value.)

If E is then given $\psi(x)$ and $d\psi(x)/dx$ can be matched at L_1 and L_2 and hence $\psi(L_3)$ can be found. The subroutine PL3FEN, which finds $\psi(L_3)$ in this way, also returns the number of nodes of $\psi(x)$. The eigenvalues of the system will correspond to those values of E for which $\psi(L_3) = 0$. The problem is therefore equivalent to the determination of the roots of the complex determinantal transcendental equation which can be set up by conventional methods. However, the method adopted here is thought to possess some pedagogical advantages. An example of the variation of $\psi(L_3)$ with energy is shown in Figure 4.

The eigenvalues are first estimated using perturbation theory. Then, with these as a guide, by repeated doubling and halving, two energies, one above and one below the true nth energy level, are found where the number of nodes are $n - 1$ and n respectively. The function EFPL30 is next employed to determine an accurate zero of $\psi(L_3)$ using a combination of the methods of linear extrapolation, linear interpolation, and bisection.

4.2 Running the program

Input the position where the perturbation starts (L_1) and ends (L_2), the width of the infinite potential well (L_3) and the mass of the particle (M_p).

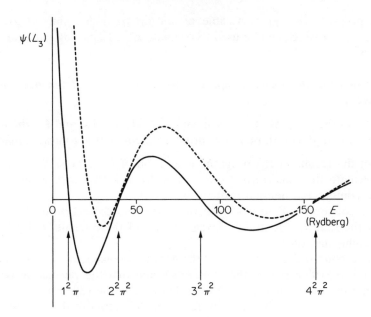

Figure 4. The variation of $\psi(L_3)$ with E, as returned by PL3FEN for $L_3 = 1.0$, $M_p = 1.0$. The solid curve shows the results for $V_p = 0$; the values of E for which $\psi(L_3) = 0$ coincide with the theoretical values for the energy: $E_n^0 = \pi^2 n^2 / M_p L_3^2 = \pi^2 n^2$. The dashed curve shows the result for $L_1 = 0.49$, $L_2 = 0.51$, $V_p = 500.0$. As we would expect from first-order perturbation theory, the ground state and second excited state energies are modified much more than the energies of the first and third excited states.

Rather than calculating the energy at a single value of V_p the program allows a range of values of V_p to be entered by reading the lowest required value of V_p (VP1)—which can be ≤ 0 or > 0, the number of different values of V_p required (NVP) and the amount by which V_p is to be increased (DVP). Choose these parameters so that the case $V_p = 0$ is included.

An approximate graph of the results can be plotted on a line printer; the last input should be YES or NO as required.

The computer will set out the results for the four lowest energies E_1, E_2, E_3, and E_4 as a function of V_p, in a table.

Remembering that the point of the exercise is to compare these computed results with perturbation theory, check that the chosen range of V_p is such that the energies change significantly (so that you can make the comparison) but not drastically (or otherwise V_p will be too large for perturbation theory to apply).

The program also gives a table of $\sin(n\pi L_1/L_3)$ and $\sin(n\pi L_2/L_3)$ for $n = 1, 2, \ldots 14$ which can be used to evaluate $E_n^{(1)}$, $E_n^{(2)}$ using equations (24) and (26).

4.3 Comparing the exact computer results with those from perturbation theory

Having run the program for chosen values of M_p, L_1, L_2, and L_3, the results can be compared with those of perturbation theory in the following manner:

(1) Plot the results of the program as a graph of E_n against V_p.
(2) Calculate the unperturbed energies and check that they coincide with the computer-generated results for $V_p = 0$.
(3) Use equation (24) to work out the first-order correction to the energy for the four lowest straight levels. A plot of these results will give four straight lines on your graph.
(4) The second-order correction to the energy is considerably more complicated to evaluate. It is therefore suggested that the calculation only be made for the ground state, i.e. evaluate equation (26) for $n = 1$, $m \neq 1$ but consider only the first few terms in the second-order correction to the energy. You will find (as is generally the case) that these decrease quite rapidly because $|E_n^0 - E_m^0|$ increases rapidly with $|n - m|$, and so the series converges quickly. Plot the second-order correction to the ground state energy on the graph.
(5) Can you make any general comments about the conditions for the validity of perturbation theory?

4.4 A Typical session

If the chosen parameters are:

L_1 (the position where the perturbation starts) $= 0.4$ ⎤
L_2 (the position where the perturbation ends) $= 0.5$ ⎬ (Bohr radii)
L_3 (the width of the infinite potential well) $= 1.0$ ⎦
M_p (the mass of the particle) $= 1.0$ (electron masses)
VP1 (the lowest value of the perturbation potential) $= -150.0$ ⎤
DVP (the amount by which the perturbing potential is to be increased) $= 10.0$ ⎬ (Rydbergs)
NVP (the required number of values of the perturbing potential) $= 31$

and an approximate graph of the results on the line printer is requested, the results of the computer program are

```
TYPE THE WIDTH OF THE INFINITE POTENTIAL WELL, I.E. L3
1.0
TYPE THE POSITION WHERE THE PERTURBATION STARTS I.E. L1
0.4
TYPE THE POSITION WHERE THE PERTURBATION ENDS I.E. L2
0.5
TYPE THE MASS OF THE PARTICLE I.E. MP
1.0
TYPE THE LOWEST VALUE OF THE PERTURBING POTENTIAL VP
-150.0
TYPE THE REQUIRED NUMBER OF VALUES OF THE PERTURBING POTENTIAL
31
TYPE THE AMOUNT BY WHICH THE PERTURBING POTENTIAL IS TO BE INCREASED
10.0
```

```
  RESULTS OF COMPUTER PROGRAM TO CALCULATE THE ENERGY OF A PARTICLE
  IN A PERTURBED INFINITE POTENTIAL WELL. THE WELL WIDTH IS  0.1000E 01
  THE PERTURBATION STARTS AT  0.4000E 00 AND ENDS AT  0.5000E 00
                 THE PARTICLE MASS IS  0.1000E 01
```

VP	E1	E2	E3	E4
-0.1500E 03	-0.3776E 02	0.3658E 02	0.7090E 02	0.1484E 03
-0.1400E 03	-0.3344E 02	0.3673E 02	0.7175E 02	0.1489E 03
-0.1300E 03	-0.2927E 02	0.3689E 02	0.7264E 02	0.1495E 03
-0.1200E 03	-0.2526E 02	0.3706E 02	0.7359E 02	0.1500E 03
-0.1100E 03	-0.2140E 02	0.3722E 02	0.7458E 02	0.1506E 03
-0.1000E 03	-0.1771E 02	0.3740E 02	0.7563E 02	0.1512E 03
-0.9000E 02	-0.1419E 02	0.3757E 02	0.7674E 02	0.1518E 03
-0.8000E 02	-0.1083E 02	0.3776E 02	0.7789E 02	0.1524E 03
-0.7000E 02	-0.7650E 01	0.3795E 02	0.7910E 02	0.1530E 03
-0.6000E 02	-0.4640E 01	0.3815E 02	0.8036E 02	0.1537E 03
-0.5000E 02	-0.1803E 01	0.3835E 02	0.8167E 02	0.1543E 03
-0.4000E 02	0.8630E 00	0.3856E 02	0.8303E 02	0.1550E 03
-0.3000E 02	0.3359E 01	0.3878E 02	0.8443E 02	0.1557E 03
-0.2000E 02	0.5690E 01	0.3901E 02	0.8586E 02	0.1564E 03
-0.1000E 02	0.7858E 01	0.3924E 02	0.8733E 02	0.1572E 03
0.0000E 00	0.9870E 01	0.3948E 02	0.8883E 02	0.1579E 03
0.1000E 02	0.1173E 02	0.3973E 02	0.9034E 02	0.1587E 03
0.2000E 02	0.1345E 02	0.3998E 02	0.9187E 02	0.1595E 03
0.3000E 02	0.1503E 02	0.4024E 02	0.9340E 02	0.1603E 03
0.4000E 02	0.1648E 02	0.4050E 02	0.9493E 02	0.1611E 03
0.5000E 02	0.1782E 02	0.4077E 02	0.9645E 02	0.1620E 03
0.6000E 02	0.1904E 02	0.4105E 02	0.9795E 02	0.1628E 03
0.7000E 02	0.2015E 02	0.4132E 02	0.9944E 02	0.1637E 03
0.8000E 02	0.2117E 02	0.4160E 02	0.1009E 03	0.1646E 03
0.9000E 02	0.2210E 02	0.4188E 02	0.1023E 03	0.1655E 03
0.1000E 03	0.2296E 02	0.4216E 02	0.1037E 03	0.1664E 03
0.1100E 03	0.2373E 02	0.4245E 02	0.1051E 03	0.1673E 03
0.1200E 03	0.2444E 02	0.4273E 02	0.1064E 03	0.1683E 03
0.1300E 03	0.2509E 02	0.4300E 02	0.1077E 03	0.1692E 03
0.1400E 03	0.2568E 02	0.4328E 02	0.1089E 03	0.1701E 03
0.1500E 03	0.2623E 02	0.4355E 02	0.1101E 03	0.1711E 03

```
DO YOU WANT TO PLOT OUT THESE RESULTS AS AN APPROXIMATE GRAPH
 ON THE TELETYPE.   TYPE YES OR NO.
YES
```

APPROXIMATE GRAPH OF THE ENERGIES OF THE GROUND STATE AND FIRST THREE
EXCITED STATES AS A FUNCTION OF THE PERTURBATION VP. VP IS PLOTTED ON
THE HORIZONTAL AXIS AND VARIES FROM -0.1500E 03 TO 0.1500E 03
THE ENERGY IS PLOTTED ON THE VERTICAL AXIS AND VARIES FROM 0.1711E 03
        TO -0.3776E 02

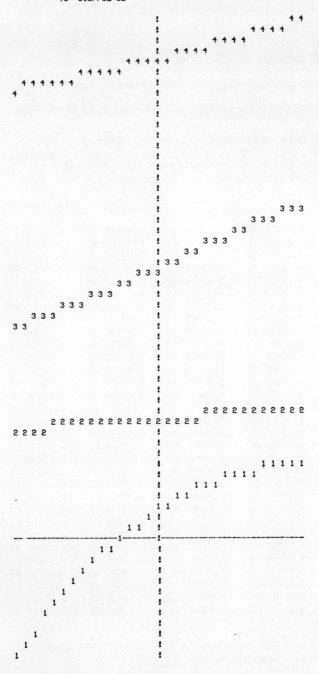

```
DO YOU WANT TO RE-RUN THE PROGRAM WITH THE SAME VALUES FOR L1, L2, L3
AND MP BUT WITH DIFFERENT VALUES FOR THE PERTURBING POTENTIAL VP.
TYPE YES OR NO
NO

    N        SIN(N*PI*L1/L3)        SIN(N*PI*L2/L3)

    1           0.951056              1.000000
    2           0.597786              0.000001
    3          -0.597784             -1.000000
    4          -0.951057             -0.000001
    5          -0.000002              1.000000
    6           0.951055              0.000004
    7           0.587788             -1.000000
    8          -0.587782             -0.000003
    9          -0.951058              1.000000
   10          -0.000004              0.000006
   11           0.951055             -1.000000
   12           0.587792             -0.000009
   13          -0.587777              1.000000
   14          -0.951059              0.000006
PROGRAM FINISHED
```

5. DISCUSSION

The unperturbed energies (in Rydbergs), for $V_p = 0$, are

$$E_n^{(0)} = \frac{n^2 \pi^2}{M_p L_3^2} \tag{29}$$

Equation (29), for the chosen parameters, gives $E_1^{(0)} = 9.87$, $E_2^{(0)} = 39.48$, $E_3^{(0)} = 88.83$, $E_4^{(0)} = 157.91$.

The first-order correction to the energy is given by equation (24). Thus, for the chosen parameters, the corrections to the first four levels are:

$n =$	1	2	3	4
$\dfrac{\text{(First-order correction)}}{V_p} =$	0.194	0.024	0.150	0.077

The second-order correction to the energy of the nth level is given by equation (25), which can be evaluated using equation (26). For the example under discussion, values of I_{mn}/V_p are given in Table 1 while $E_{mn}^{(2)}/V_p^2$ values are given in Table 2. The ranges of n and m are restricted to $1 \leqslant n \leqslant 4$ and $1 \leqslant m \leqslant 10$.

Table 1. Values of I_{mn}/V_p

	$m = 1$	2	3	4	5	6	7	8	9	10
n										
1	—	0.059	-0.169	-0.107	0.126	0.136	-0.074	-0.141	0.023	0.125
2	0.059	—	-0.048	-0.043	0.028	0.052	-0.005	-0.050	-0.016	0.039
3	-0.169	-0.048	—	0.088	-0.117	-0.113	0.076	0.120	-0.036	-0.111
4	-0.107	-0.043	0.088	—	-0.053	-0.094	0.013	0.091	0.025	-0.072

Table 2. Values of $E_{mn}^{(2)}/V_p^2$

	$m = 1$	2	3	4	5	6	7	8	9	10	Total
n											
1	—	-1.19×10^{-4}	-3.63×10^{-4}	-7.79×10^{-5}	-6.72×10^{-5}	-5.34×10^{-5}	-1.15×10^{-5}	-3.21×10^{-5}	-6.93×10^{-7}	-1.61×10^{-5}	-7.40×10^{-4}
2	1.19×10^{-4}	—	-4.69×10^{-5}	-1.57×10^{-5}	-3.91×10^{-6}	-8.66×10^{-6}	-6.43×10^{-8}	-4.30×10^{-6}	-3.26×10^{-7}	-1.60×10^{-6}	$+3.74 \times 10^{-5}$
3	3.63×10^{-4}	4.69×10^{-5}	—	-1.12×10^{-4}	-8.66×10^{-5}	-4.77×10^{-5}	-1.45×10^{-5}	-2.66×10^{-5}	-1.71×10^{-6}	-1.37×10^{-5}	$+1.07 \times 10^{-5}$
4	7.79×10^{-5}	1.57×10^{-5}	1.12×10^{-4}	—	-3.21×10^{-5}	-4.43×10^{-5}	-4.97×10^{-7}	-1.76×10^{-5}	-9.58×10^{-7}	-6.27×10^{-7}	$+1.03 \times 10^{-4}$

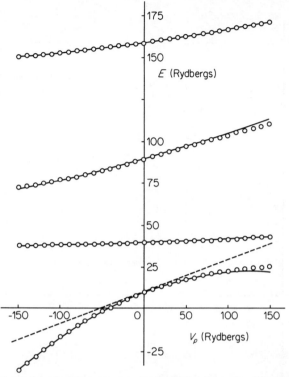

Figure 5. The results of the computer program (circles) for
$L_1 = 0.4$, $L_2 = 0.5$, $L_3 = 1.0$, $M_p = 1.0$ compared with the perturbation theory calculation. The solid curves are the results for
first- plus second-order perturbation theory. The dashed line for
the ground state shows the result for first-order perturbation
theory alone

Figure 5 shows a comparison between the results of the computer program (the circles) and perturbation theory. The solid curves are the results for
first- plus second-order perturbation theory corrections. The dashed line for
the ground state shows the result for first-order perturbation theory alone. It
can be seen that, even with the inclusion of the second-order correction, the
results are starting to deviate from the (more accurate) computer results at
the largest values of $|V_p|$. For these values of V_p the amount of admixture of
other states into the unperturbed eigenstate is not small—which is the
situation in which we would indeed expect second-order perturbation theory
to break down.

Though they correspond to situations far outside the range of validity of
perturbation theory the results for $V_p \gg E_n^0$ and $V_p \to -\infty$ are of some
interest.

For $V_p \gg E_n^0$ the situation tends towards a pair of isolated infinite potential

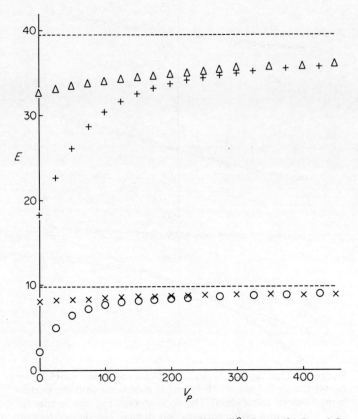

Figure 6. The energy levels for $V_p \gg E_n^0$. $L_1 = 1.0$, $L_2 = 1.2$, $L_3 = 2.2$, $M_p = 1.0$. The dashed lines indicate the ground state and first excited state energies for the particle in an infinite potential well of width $L_1 = L_3 - L_2$

wells, of width L_1 and $L_3 - L_2$. It will be found for this case that the energies tend towards the (constant) values for the infinite potential wells with these widths. Furthermore, if $L_1 = L_3 - L_2$ these levels are degenerate. This kind of situation is shown in Figure 6.

For $V_p \rightarrow -\infty$ the situation tends towards an infinite potential well of width $L_2 - L_1$ with the zero of energy shifted to V_p. Hence

$$E_n \rightarrow V_p + \frac{n^2 \pi^2}{M_p (L_2 - L_1)^2}.$$

An example of this behaviour is shown in Figure 7.

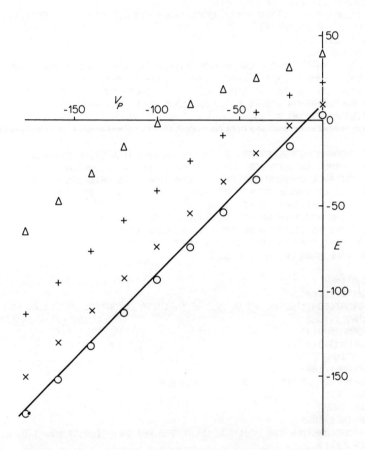

Figure 7. The energy levels for $V_p \to -\infty$ and $L_1 = 0.5$, $L_2 = 1.5$, $L_3 = 2.0$, $M_p = 1.0$. The straight line corresponds to $E = V_p + \pi^2/M_p(L_2 - L_1)^2$ and it can be seen that the ground state energy tends asymptotically to this value for $V_p \to -\infty$

MAIN PROGRAM

```
C  THIS PROGRAM USES AN ITERATIVE TECHNIQUE TO FIND THE GROUND STATE AND
C   FIRST THREE EXCITED STATES OF A PARTICLE IN AN INFINITE POTENTIAL
C   WELL WITH AN ADDITIONAL PERTURBING POTENTIAL IN PART OF
C   THE WELL.
       COMMON L1,L2,L3,VP,MP,N,PI
       DIMENSION SL1(15),SL2(15),HDV(5,10),EV(10),EC2(5),EP(5),E(51,4),
      1 PLOT(120),NAME(40)
       REAL L1,L2,L3,MP
       LOGICAL BOOL
       DATA Y,YN,PLOT1,PLOT2,PLOT3,PLOT4,PLOT5,PLOT6,PLOT7/4HYES ,4HNO  ,
      1 4H1  ,4H2  ,4H3  ,4H4  ,4H  ,4H- ,4H! /
C  THE WIDTH OF THE INFINITE POTENTIAL WELL (=L3) AND THE POSITION WHERE
C   THE PERTURBATION STARTS (=L1) AND ENDS (=L2) ARE READ IN.
       PI=ATAN(1.0)*4.0
C*********************************************************************************
C
C   THE PROGRAM AS PRESENTED IS FOR USE ON AN INTERACTIVE SYSTEM.
C   FREE FORMAT IS USED FOR INPUT  - F0.0 AND I0  -  WHICH MAY NOT
C   BE AVAILABLE ON SOME SYSTEMS.  ACCORDINGLY THE MATERIAL UP TO THE
C   NEXT SET OF ******'S MAY NEED SOME MODIFICATION.  IT IS NECESSARY
C   TO READ IN AT THIS POINT IN THE PROGRAM:
C   L1 (THE POSITION WHERE THE PERTURBATION STARTS)
C   L2 (THE POSITION WHERE THE PERTURBATION ENDS)
C   L3 (THE WIDTH OF THE INFINITE POTENTIAL WELL)
C   MP (THE MASS OF THE PARTICLE)
C
     2 CONTINUE
       WRITE(1,100)
   100 FORMAT(53HTYPE THE WIDTH OF THE INFINITE POTENTIAL WELL,I.E. L3)
       READ(1,1)L3
     1 FORMAT(F0.0)
       IF(L3)3,3,4
     3 CONTINUE
       WRITE(1,102)
   102 FORMAT(27HL3 MUST BE GREATER THAN 0.0)
       GOTO 2
     4 CONTINUE
       WRITE(1,106)
   100 FORMAT(55HTYPE THE POSITION WHERE THE PERTURBATION STARTS I.E. L1)
       READ(1,1)L1
       IF(L1.GE.0.0.AND.L1.LT.L3)GOTO 6
       WRITE(1,108)
   108 FORMAT(30HL1 MUST LIE BETWEEN 0.0 AND L3)
       GOTO 4
     6 CONTINUE
       WRITE(1,110)
   110 FORMAT(53HTYPE THE POSITION WHERE THE PERTURBATION ENDS I.E. L2)
       READ(1,1)L2
       IF(L2.GT.L1.AND.L2.LE.L3)GOTO 8
       WRITE(1,112)
   112 FORMAT(29HL2 MUST LIE BETWEEN L1 AND L3)
       GOTO 6
     8 CONTINUE
C NEXT THE MASS OF THE PARTICLE (=MP) IS INPUT.
       WRITE(1,114)
   114 FORMAT(37HTYPE THE MASS OF THE PARTICLE I.E. MP)
```

```
MAIN PROGRAM
      READ(1,1)MP
      IF(MP)9,9,10
    9 CONTINUE
      WRITE(1,116)
  116 FORMAT(49HTHE MASS OF THE PARTICLE MUST BE GREATER THAN 0.0)
      GOTO 8
   10 CONTINUE
C*********************************************************************************
C AT THIS POINT IN THE PROGRAM THE MAIN TERMS IN THE FIRST AND SECOND
C   ORDER PERTURBATION THEORY CORRECTIONS TO THE ENERGIES OF THE 5
C   LOWEST ENERGY STATES ARE COMPUTED. THEY ARE USED TO GUIDE THE
C   SUBSEQUENT ITERATIVE SEARCH FOR THE EIGENVALUES. THE RESULTS ARE
C   STORED AND SOME OF THEM ARE OUTPUT AT THE END OF THE PROGRAM AS
C   AN AID TO THE STUDENT.
      DO 12 I=1,15
      SL1(I)=SIN(FLOAT(I)*PI*L1/L3)
      SL2(I)=SIN(FLOAT(I)*PI*L2/L3)
   12 CONTINUE
      DO 14 N=1,5
      HDV(N,N)=(L2-L1)/L3-(SL2(2*N)-SL1(2*N))/(2.0*PI*FLOAT(N))
      NP=N+1
      DO 16 M=NP,10
      MMN=M-N
      MPN=M+N
      HDV(N,M)=(SL2(MMN)-SL1(MMN))/FLOAT(MMN)
      HDV(N,M)=(HDV(N,M)-(SL2(MPN)-SL1(MPN))/FLOAT(MPN))/PI
   16 CONTINUE
   14 CONTINUE
      DO 18 N=2,5
      NM=N-1
      DO 20 M=1,NM
      HDV(N,M)=HDV(M,N)
   20 CONTINUE
   18 CONTINUE
C COMPUTE THE UNPERTURBED ENERGIES (=EU(N))
      DO 22 N=1,10
      EU(N)=PI*PI*FLOAT(N)*FLOAT(N)/(MP*L3*L3)
   22 CONTINUE
C AS YET THE PERTURBING POTENTIAL HAS NOT BEEN READ IN. HOWEVER,
C   WE WISH TO BE ABLE TO EXPRESS THE ENERGY, UP TO SECOND ORDER, AS:
C             E = EU(N) + VP*HDV(N,N) + VP*VP*EC2(N) .   THUS.
      DO 24 N=1,5
      EC2(N)=0.0
      DO 26 M=1,10
      IF(M-N)25,26,25
   25 CONTINUE
      EC2(N)=EC2(N)+(HDV(N,M)*HDV(N,M))/(EU(N)-EU(M))
   26 CONTINUE
   24 CONTINUE
   28 CONTINUE
C THE VALUES OF THE PERTURBATION (=VP) ARE READ IN
C*********************************************************************************
C
C    FURTHER INPUT IS REQUIRED HERE
C    VP1 (THE LOWEST VALUE OF THE PERTURBING POTENTIAL VP)
C    DVP (THE AMOUNT BY WHICH THE PERTURBING POTENTIAL IS TO BE INCREASED)
```

MAIN PROGRAM

```
C     NVP (THE REQUIRED NUMBER OF VALUES OF THE PERTURBING POTENTIAL)
C          NOTE THAT NVP IS AN INTEGER VARIABLE  -  THE OTERS ARE REAL
C
      WRITE(1,118)
  118 FORMAT(52HTYPE THE LOWEST VALUE OF THE PERTURBING POTENTIAL VP)
      READ(1,1)VP1
   30 CONTINUE
      WRITE(1,120)
  120 FORMAT(52HTYPE THE REQUIRED NUMBER OF VALUES OF THE PERTURBING ,
     1 9HPOTENTIAL)
      READ(1,29)NVP
   29 FORMAT(I0)
      IF(NVP.GT.0.AND.NVP.LT.52)GOTO 32
      WRITE(1,122)
  122 FORMAT(53HTHE REQUIRED NUMBER MUST BE POSITIVE AND LESS THAN 52)
      GOTO 30
   32 CONTINUE
      WRITE(1,124)
  124 FORMAT(55HTYPE THE AMOUNT BY WHICH THE PERTURBING POTENTIAL IS TO ,
     1 12HBE INCREASED)
      READ(1,1)DVP
      WRITE(1,208)
      WRITE(1,208)
C********************************************************************************
      WRITE(1,200)
  200 FORMAT(1H ,45HRESULTS OF COMPUTER PROGRAM TO CALCULATE THE ,
     1 20HENERGY OF A PARTICLE)
      WRITE(1,202)L3
  202 FORMAT(1H ,49HIN A PERTURBED INFINITE POTENTIAL WELL. THE WELL ,
     1 9HWIDTH IS ,E11.4)
      WRITE(1,204)L1,L2
  204 FORMAT(1H ,27HTHE PERTURBATION STARTS AT ,E11.4,13H AND ENDS AT ,
     1 E11.4)
      WRITE(1,206)MP
  206 FORMAT(1H ,15X,21HTHE PARTICLE MASS IS ,E11.4)
C SET OUT THE HEADING FOR THE RESULTS
      WRITE(1,208)
  208 FORMAT(1H )
      WRITE(1,210)
  210 FORMAT(1H ,6X,2HVP,12X,2HE1,12X,2HE2,12X,2HE3,12X,2HE4)
      WRITE(1,208)
      DO 34 I=1,NVP
      VP=VP1+FLOAT(I-1)*DVP
C COMPUTE THE ENERGY OF THE 5 LOWEST STATES USING 2ND ORDER PERTURBATION
      DO 36 N=1,5
      EP(N)=EU(N)+VP*HDV(N,N)+VP*VP*EC2(N)
   36 CONTINUE
C SORT THE EP(N) INCASE THE PERTURBATION HAS PRODUCED A CROSS OVER
C    OF THE LEVELS
   39 CONTINUE
      DO 37 N=1,4
      IF(EP(N).LT.EP(N+1))GOTO 37
      EPN=EP(N+1)
      EPNP=EP(N)
      EP(N)=EPN
      EP(N+1)=EPNP
```

MAIN PROGRAM

```
   37 CONTINUE
      IF(EP(1).GT.EP(2).OR.EP(2).GT.EP(3).OR.EP(3).GT.EP(4).
     1 OR.EP(4).GT.EP(5))GOTO 39
C   THE PROGRAM NOW STARTS TO SEARCH FOR VALUES OF ENERGY SUCH THAT
C   THE WAVE FUNCTION VANISHES AT L3.
C   IN THE FIRST PART OF THE SEARCH FOR THE N'TH EIGENSTATE BY
C   USING A BISECTION TECHNIQUE THE PPOGRAM FINDS TWO ENERGIES, ONE ABOVE
C   AND ONE BELOW THE TRUE N'TH ENERGY LEVEL, WHERE THE NUMBER OF NODES
C   ARE N-1 AND N RESPECTIVELY. THESE ENERGIES ARE THEN USED AS STARTING
C   POINTS FOR THE LINEAR INTERPOLATION ROOT FINDING FUNCTION.
      EF=AMIN1(0.0,VP)
      CALL PL3FEN(EF,PL3F,NODES)
      DO 38 N=1,4
      ET=(EP(N)+EP(N+1))*0.5
      GAP=ABS(ET-EF)
      IF(GAP.EQ.0.0)GAP=EU(1)
      ET=EF+GAP
      BOOL=.FALSE.
      DO 40 J=1,500
      CALL PL3FEN(ET,PL3T,NODES)
      PL3S=PL3T
      IF(NODES.GT.N)GOTO 76
      IF(NODES.LT.N)GOTO 78
      ES=ET
      GOTO 44
   76 CONTINUE
      BOOL=.TRUE.
      ET=ET-GAP
      GOTO 42
   78 CONTINUE
      ET=ET+GAP
   42 CONTINUE
      IF(BOOL)GAP=GAP*0.5
      IF(.NOT.BOOL)GAP=GAP*2.0
   40 CONTINUE
      WRITE(1,214)N
  214 FORMAT(1H0,49H WARNING - THE PROGRAM FAILED TO FIND A CORRECT  ,
     1 9HSTARTING ,/,26HPOINT FOR THE SEARCH WITH ,I4,6H NODES)
      ES=ET
   44 CONTINUE
C THE ACCURACY FOUND BY EFPL30 IS SET BY THE PARAMETERS EAC AND FAC.
      EAC=EU(1)*1.0E-5
      FAC=1.0E-5
      E(I,N)=EFPL30(EF,ES,PL3F,PL3S,EAC,FAC)
      IF(N-4)45,38,38
   45 CONTINUE
      EF=ET
      PL3F=PL3T
   38 CONTINUE
      WRITE(1,130)VP,E(I,1),E(I,2),E(I,3),E(I,4)
  130 FORMAT(1H ,5(E11.4,3X))
   34 CONTINUE
C   THE OPTION IS NOW PROVIDED OF PLOTTING A GRAPH OF E AGAINST VP ON THE
C     PRINTER.
C************************************************************************
C
```

MAIN PROGRAM

```
C     THE OPTION IS PROVIDED OF PLOTTING A GRAPH ON THE LINE PRINTER.
C     IF   YES   OR   NO   IS ENTERD AS REQUIRED IT CAN BE READ IN A
C     FORMAT AND COMPARED WITHE THE VARIABLES Y(=4HYES ,) AND
C     YN(=4HNO  ,) SET IN THE DATA STATEMENT. IF A GRAPH IS NOT REQUIRED
C     CONTROL CAN BE TRANSFERED TO STATEMENT 62
C
      WRITE(1,208)
      WRITE(1,170)
  170 FORMAT(//44HDO YOU WANT TO PLOT OUT THESE RESULTS AS AN,
     1 17HAPPROXIMATE GRAPH,/35H ON THE TELETYPE.   TYPE YES OR NO.)
   64 CONTINUE
      READ(1,171)YON
  171 FORMAT(A4)
      IF(YON.EQ.Y)GOTO 60
      IF(YON.EQ.YN)GOTO 62
      WRITE(1,174)
  174 FORMAT(1H ,40HRESPONSE NOT RECOGNISED.  TYPE YES OR NO)
      GOTO 64
   60 CONTINUE
C*************************************************************************
C     THE MAXIMUM WIDTH OF THE GRAPH IS SET BY THE PARAMETER IWG. THIS
C     IS CHOSEN SUCH THAT IWG-1 IS A MULTIPLE OF NVP-1
      IWG=70
      GAP=FLOAT(IWG-1)/FLOAT(NVP-1)
      IGAP=IFIX(AINT(GAP))
      IWG=(NVP-1)*IGAP+1
      DDVP=ABS(DVP/FLOAT(IGAP))
C     FIND THE MAXIMUM AND MINIMUM VALUES OF THE ENERGY
      EMIN=E(1,1)
      EMAX=E(1,1)
      DO 66 N=1,4
      DO 68 I=2,NVP
      IF(E(I,N).LT.EMIN)EMIN=E(I,N)
      IF(E(I,N).GT.EMAX)EMAX=E(I,N)
   68 CONTINUE
   66 CONTINUE
      IF(EMIN.GT.0.0)EMIN=0.0
C     THE HEIGHT OF THE GRAPH IS SET BY THE PARAMETER IHG
      IHG=IWG
      SCALE=(EMAX-EMIN)/FLOAT(IHG-1)
C     WRITE OUT HEADING FOR THE GRAPH
      VP2=VP1+FLOAT(NVP-1)*DVP
      WRITE(1,208)
      WRITE(1,218)
  218 FORMAT(1H ,48HAPPROXIMATE GRAPH OF THE ENERGIES OF THE GROUND ,
     1 21HSTATE AND FIRST THREE)
      WRITE(1,220)
  220 FORMAT(1H ,49HEXCITED STATES AS A FUNCTION OF THE PERTURBATION ,
     1 20HVP. VP IS PLOTTED ON)
      WRITE(1,222)VP1,VP2
  222 FORMAT(1H ,36HTHE HORIZONTAL AXIS AND VARIES FROM ,E11.4,4H TO ,
     1 E11.4)
      WRITE(1,224)EMAX
  224 FORMAT(1H ,47HTHE ENERGY IS PLOTTED ON THE VERTICAL AXIS AND ,
     1 12HVARIES FROM ,E11.4)
      WRITE(1,226)EMIN
```

MAIN PROGRAM

```
  226 FORMAT(1H ,10X,3HTO ,E11.4)
      WRITE(1,208)
C PLOT GRAPH
      DO 70 J=1,IHG
      ETS=EMAX-SCALE*(FLOAT(J)-1.5)
      EBS=EMAX-SCALE*(FLOAT(J)-0.5)
      DO 72 I=1,NVP
      K1=(I-1)*IGAP+1
      K2=I*IGAP
      DO 74 K=K1,K2
      PLOT(K)=PLOT5
      IF(ETS.GE.0.0.AND.EBS.LT.0.0)PLOT(K)=PLOT6
      VP=VP1+DVP*FLOAT(K-1)/FLOAT(IGAP)
      IF(VP.LE.(DDVP/2.0).AND.VP.GT.(-DDVP/2.0))PLOT(K)=PLOT7
   74 CONTINUE
      IF(E(I,1).LE.ETS.AND.E(I,1).GT.EBS)PLOT(K1)=PLOT1
      IF(E(I,2).LE.ETS.AND.E(I,2).GT.EBS)PLOT(K1)=PLOT2
      IF(E(I,3).LE.ETS.AND.E(I,3).GT.EBS)PLOT(K1)=PLOT3
      IF(E(I,4).LE.ETS.AND.E(I,4).GT.EBS)PLOT(K1)=PLOT4
   72 CONTINUE
      WRITE(1,178)(PLOT(K),K=1,IWG)
  178 FORMAT(1H ,120A1)
   70 CONTINUE
   62 CONTINUE
C*****************************************************************************
C
C     THE OPTION IS NOW PROVIDED  - WITH AN INTERACTIVE SYSTEM - OF TRYING
C          DIFFERENT SETS OF VP'S.
C
      WRITE(1,132)
  132 FORMAT(/54HDO YOU WANT TO RE-RUN THE PROGRAM WITH THE SAME VALUES,
     1 14HFOR L1, L2, L3,/41HAND MP BUT WITH DIFFERENT VALUES FOR THE ,
     2 24HPERTURBING POTENTIAL VP.,/15H TYPE YES OR NO)
   46 CONTINUE
      READ(1,171)YON
      IF(YON.EQ.Y)GOTO 28
      IF(YON.EQ.YN)GOTO 48
      WRITE(1,136)
  136 FORMAT(39HRESPONSE NOT RECOGNISED. TYPE YES OR NO)
      GOTO 46
   48 CONTINUE
C*****************************************************************************
C OUTPUT SUPPLEMENTARY INFORMATION TO AID STUDENT'S PERTURBATION
C    CALCULATION.
      WRITE(1,208)
      WRITE(1,228)
  228 FORMAT(1H ,5X,1HN,6X,15HSIN(N*PI*L1/L3),6X,15HSIN(N*PI*L2/L3))
      WRITE(1,208)
      WRITE(1,140)(I,SL1(I),SL2(I),I=1,14)
  140 FORMAT(4X,I2,8X,F10.6,11X,F10.6)
      WRITE(1,312)
  312 FORMAT(1H ,16HPROGRAM FINISHED)
      STOP
      END
```

```
SUBROUTINE PL3FE
      SUBROUTINE PL3FE(E,PL3FE,NODES)
C  SUBROUTINE TO RETURN THE VALUE OF THE WAVE FUNCTION AT L3
C   AND THE NUMBER OF NODES BETWEEN 0 AND L3
C   GIVEN THE ENERGY E, THE PERTURBATION VP AND THE DIMENSIONS.
C   THE DIMENSIONS (=L1,L2,L3), PERTURBING POTENTIAL (=VP) AND
C   PARTICLE MASS (=MP) ARE ALL TRANSFERED IN THE COMMON BLOCK.
C   THE MNEMONICS USED FOR THESE ARE AS IN THE MAIN PROGRAM.
      COMMON L1,L2,L3,VP,MP,N,PI
      REAL L1,L2,L3,MP,K1,K2
C  IF VP = 0 THE CALCULATION OF THE WAVE FUNCTION AT
C   L3 IS SIMPLE.
      IF(VP.NE.0.0)GOTO 1
      IF(E)5,7,3
    7 CONTINUE
      PL3FE=L3
      NODES=0
      GOTO 24
    3 CONTINUE
      K1=SQRT(MP*E)
      PL3FE=SIN(K1*L3)/K1
      NODES=IFIX(L3*K1/PI)
      GOTO 24
    5 CONTINUE
      NODES=0
      ALPHA=SQRT(-MP*E)
      F1=EXP(ALPHA*L3)
      F2=1.0/F1
      PL3FE=(F1-F2)/(2.0*ALPHA)
      GOTO 24
    1 CONTINUE
      NODES=0
C  CALCULATE THE WAVE FUNCTION AT L1 (=PL1) AND ITS SLOPE AT L1 (=DPL1)
      IF(E)2,4,6
    2 CONTINUE
      ALPHA=SQRT(-MP*E)
      F1=EXP(ALPHA*L1)
      F2=1.0/F1
      PL1=(F1-F2)/(2.0*ALPHA)
      DPL1=(F1+F2)/2.0
      GOTO 8
    4 CONTINUE
      PL1=L1
      DPL1=1.0
      GOTO 8
    6 CONTINUE
      K1=SQRT(MP*E)
      PL1=SIN(K1*L1)/K1
      DPL1=COS(K1*L1)
      NODES=IFIX(L1*K1/PI)
    8 CONTINUE
C  CALCULATE THE WAVE FUNCTION AT L2 (=PL2) AND ITS SLOPE AT L2 (=DPL2)
      IF(E-VP)10,12,14
   10 CONTINUE
      BETA=SQRT(-MP*(E-VP))
      F1=EXP(BETA*(L1-L2))
      A=PL1-DPL1/BETA
```

```
SUBROUTINE PL3FE
      B=PL1+DPL1/BETA
      PL2=A*F1/2.0+B/(2.0*F1)
      DPL2=BETA*(-A*F1/2.0+B/(2.0*F1))
      IF(PL1.EQ.0.0)GOTO 16
      IF(PL2.NE.0.0)GOTO 11
      NODES=NODES+1
      GOTO 16
   11 CONTINUE
      IF((PL1/PL2).LT.0.0)NODES=NODES+1
      GOTO 16
   12 CONTINUE
      A=DPL1
      B=PL1-A*L1
      PL2=A*L2+B
      DPL2=A
      IF(PL1.EQ.0.0)GOTO 16
      IF(PL2.NE.0.0)GOTO 13
      NODES=NODES+1
      GOTO 16
   13 CONTINUE
      IF((PL1/PL2).LT.0.0)NODES=NODES+1
      GOTO 16
   14 CONTINUE
      K2=SQRT(MP*(E-VP))
      IF(L1.NE.0.0)GOTO 15
      PL2=SIN(K2*L2)/K2
      DPL2=COS(K2*L2)
      NODES=NODES+IFIX(L2*K2/PI)
      GOTO 16
   15 CONTINUE
      DELTA=ATAN(K2*PL1/DPL1)-K2*L1
      A=PL1/SIN(K2*L1+DELTA)
      PL2=A*SIN(K2*L2+DELTA)
      DPL2=K2*A*COS(K2*L2+DELTA)
      NODES=NODES+IFIX((L2*K2+DELTA)/PI+1.0)-IFIX((L1*K2+DELTA)/PI+1.0)
   16 CONTINUE
C  CALCULATE THE WAVE FUNCTION AT L3 (=PL3FE)
      IF(E)18,20,22
   18 CONTINUE
      F1=EXP(ALPHA*L2)
      B=(PL2+DPL2/ALPHA)/(2.0*F1)
      A=PL2*F1-B*F1*F1
      F1=EXP(ALPHA*L3)
      PL3FE=A/F1+B*F1
      IF(PL2.EQ.0.0)GOTO 24
      IF(PL3FE.NE.0.0)GOTO 17
      NODES=NODES+1
      GOTO 24
   17 CONTINUE
      IF((PL2/PL3FE).LT.0.0)NODES=NODES+1
      GOTO 24
   20 CONTINUE
      A=DPL2
      B=PL2-A*L2
      PL3FE=A*L3+B
      IF(PL2.EQ.0.0)GOTO 24
```

```
SUBROUTINE PL3FE
      IF(PL3FE.NE.0.0)GOTO 19
      NODES=NODES+1
      GOTO 24
   19 CONTINUE
      IF((PL2/PL3FE).LT.0.0)NODES=NODES+1
      GOTO 24
   22 CONTINUE
      IF(L2.NE.L3)GOTO 26
      PL3FE=PL2
      GOTO 24
   26 CONTINUE
      DELTA=ATAN(K1*PL2/DPL2)-K1*L2
      A=PL2/SIN(K1*L2+DELTA)
      PL3FE=A*SIN(K1*L3+DELTA)
      NODES=NODES+IFIX((L3*K1+DELTA)/PI+1.0)-IFIX((L2*K1+DELTA)/PI+1.0)
   24 CONTINUE
      RETURN
      END
```

```
FUNCTION EFPL30
      FUNCTION EFPL30(EF,ES,PL3F,PL3S,EAC,FAC)
C   THIS FUNCTION LOCATES THE VALUE OF THE ENERGY =EFPL30 WHEN THE WAVE
C    FUNCTION AT L3 IS ZERO IN THE INTERVAL EF TO ES BY A COMBINATION OF
C    THE METHODS OF LINEAR EXTRAPOLATION, INTERPOLATION AND BISECTION.
C    THE ACCURACY IS SPECIFIED BY THE PARAMETERS EAC AND FAC.
      COMMON L1,L2,L3,VP,MP,N,PI
      REAL L1,L2,L3,MP
      LOGICAL SWITCH
      SWITCH=.FALSE.
      EA=EF
      EB=ES
      PL3A=PL3F
      PL3B=PL3S
      EX=PL3S*(ES-EF)/(PL3S-PL3F)
      ET=ES-EX
      DO 2 I=1,1000
      CALL PL3FEN(ET,PL3T,NODES)
      IF(ABS(PL3T).GT.(L3*FAC))GOTO 16
      EFPL30=ET
      GOTO 14
   16 CONTINUE
C   TEST IF ET LIES OUTSIDE PREVIOSLY FOUND VALUES
      IF(ET.GE.EB.OR.ET.LE.EA)GOTO 4
C   TEST WHETHER EXTRAPOLATION WOULD INVOLVE DIVISION BY 0
      IF(PL3T.EQ.PL3S)GOTO 4
C   RESET EA AND EB
      IF(NODES.EQ.N)GOTO 10
      EA=ET
      PL3A=PL3T
      GOTO 12
   10 CONTINUE
      EB=ET
      PL3B=PL3T
   12 CONTINUE
C   USE LINEAR EXTRAPOLATION
      EF=ES
      PL3F=PL3S
      ES=ET
      PL3S=PL3T
      EX=PL3S*(ES-EF)/(PL3S-PL3F)
      IF((ABS(EX)).LT.EAC)GOTO 6
      IF(ET.EQ.0.0)GOTO 18
      IF(ABS(EX/ET).LT.FAC)GOTO 6
   18 CONTINUE
      ET=ES-EX
      GOTO 2
    4 CONTINUE
      IF(SWITCH)GOTO 8
C   USE LINEAR INTERPOLATION FROM THE TWO CLOSEST PREVIOUS VALUES
      EX=PL3A*(EA-EB)/(PL3A-PL3B)
      ET=EA-EX
      SWITCH=.TRUE.
      IF(ABS(EX).GT.EAC)GOTO 2
      IF(ET.EQ.0.0)GOTO 2
      IF(ABS(EX/ET).GT.FAC)GOTO 2
   20 CONTINUE
```

```
FUNCTION EFPL30

      EFPL30=ET
      GOTO 14
   8 CONTINUE
C   ALTERNATIVELY, USE BISECTION FROM TWO CLOSEST PREVIOUS VALUES
      EX=(EA-EB)*0.5
      ET=(EA+EB)*0.5
      SWITCH=.FALSE.
      IF(ABS(EX).GT.EAC)GOTO 2
      IF(ET.EQ.0.0)GOTO 22
      IF(ABS(EX/ET).LT.FAC)GOTO 6
  22 CONTINUE
      EFPL30=ET
      GOTO 14
   2 CONTINUE
      WRITE(1,100)N
 100 FORMAT(1H ,44HWARNING : PROGRAM DOES NOT FIND THE CORRECT ,I1,
    1 3H TH)
      WRITE(1,102)VP
 102 FORMAT(1H ,20HEIGENSTATE FOR VP = ,E12.5)
      EFPL30=ES
      GOTO 14
   6 CONTINUE
      EFPL30=ES-EX
  14 CONTINUE
      RETURN
      END
```

Physics Programs
Edited by A. D. Boardman
© 1980 John Wiley & Sons Ltd.

CHAPTER 9

Simulation of Phonon Dispersion Curves and Density of States

G. J. KEELER

1. INTRODUCTION

In order to study the vibrational properties of crystalline solids, it is necessary to know in detail the frequency dependence of the normal modes of vibration of the crystal lattice.

Neutron-scattering measurements and other experimental observations provide overwhelming evidence that the normal modes are quantized, with energy $\hbar\omega$, and these quantized vibrations are referred to as phonons. An understanding of both the microscopic properties of the phonons, and macroscopic properties related to thermal vibrations (such as the specific heat and optical properties of insulating materials), requires a knowledge of the phonon dispersion curves and density of states.

In spite of the fundamental role played by the density states, it is very difficult to measure directly, and is almost invariably computed. Experimental data will normally give information on the phonon dispersion curves, and these will then be used to determine the interatomic force constants, by comparing the experimental curves with those calculated from the force constants. Even when these have been determined, it is by no means simple to calculate the density of states analytically, but modern computing methods have proved an ideal tool for solving the problem numerically.

2. LINEAR ATOMIC CHAIN

2.1 Dispersion curves

Before discussing dispersion curves in detail, it is worth pointing out that although calculation of specific heats, for example, requires a quantum-mechanical treatment, phonon dispersion curves can be calculated from a purely classical treatment.

Many of the salient features of lattice vibrations can be most clearly illustrated by consideration of a linear chain of atoms, as shown in Figure 1.

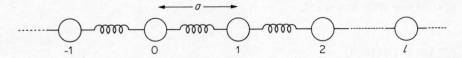

Figure 1. Linear chain of identical atoms, shown schematically with interatomic forces between nearest neighbours only.

We suppose that there are interactions between each pair of atoms in the chain (although Figure 1 has for simplicity been drawn to suggest forces between nearest neighbours only). We shall assume that the forces obey Hooke's law, i.e. they are linear in the relative displacements of the atoms (the 'harmonic approximation'), and we can then define interatomic force constants γ_l such that if u_l denotes the displacement of the lth atom, then the force on the atom at the origin due to the displacement u_l is

$$F_0 = +\gamma_l(u_l - u_0). \tag{1}$$

Summing over all atoms (l +ve and −ve), the equation of motion of the atom at the origin is

$$m\frac{\partial^2 u_0}{\partial t^2} = \sum_{l \neq 0} \gamma_l(u_l - u_0). \tag{2}$$

If we look for a solution for u_0 in the form of a wave of frequency ω and wave number $k(= 2\pi/\lambda)$ travelling in the x-direction, this will have the general form

$$u(x) = A\,e^{i(\omega t - kx)}. \tag{3}$$

However, we need only consider displacements at actual atomic sites, so if the lattice spacing is a,

$$u_l = A\,e^{i(\omega t - kla)},$$

giving

$$-m\omega^2 A e^{i\omega t} = \sum_{l \neq 0} \gamma_l A\{e^{i(\omega t - kla)} - e^{i\omega t}\}. \tag{4}$$

Since by symmetry $\gamma_l = \gamma_{-l}$, we may rewrite this, after cancellation, as

$$m\omega^2 = \sum_{l > 0} 2\gamma_l(1 - \cos kla). \tag{5}$$

Consider the simple example of nearest-neighbour interactions only ($\gamma_1 = \gamma$, $\gamma_2 = \gamma_3 = \ldots = 0$). Then equation (5) gives

$$\omega^2(k) = \frac{2\gamma}{m}(1 - \cos ka), \quad \text{or} \quad \omega(k) = 2\sqrt{\frac{\gamma}{m}}\,|\sin \tfrac{1}{2}ka|. \tag{6}$$

Thus the $\omega-k$ relationship, which we call the *dispersion curve*, is periodic in k as shown in Figure 2. However, let us consider the physical significance of this periodicity by looking at the relative motion of two successive atoms:

$$\frac{u_1}{u_0} = \frac{A\,e^{i(\omega t - ka)}}{A\,e^{i\omega t}} = e^{-ika}. \tag{7}$$

Thus a range of values of k of $2\pi/a$ covers all possible values of u_1/u_0. Since k must be allowed both positive and negative values to represent waves propagating in either direction, the range of independent values of k is

$$-\frac{\pi}{a} \leqslant k \leqslant \frac{\pi}{a}, \tag{8}$$

and this is called the *first Brillouin zone.*

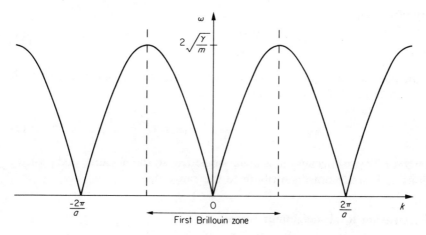

Figure 2. Plot of ω versus k (phonon dispersion curve) for a linear chain of atoms with interactions between nearest neighbours only.

2.2 Density of states

The above treatment assumes that all values of k are possible. However, physically the chain must be bounded, so that we must apply appropriate boundary conditions. The particular choice of boundary condition affects only the fine details of the density of states, and it is sufficient to consider the most commonly utilized boundary condition, which is for the solution to be periodic over a very large number of lattice spacings, N. Thus $u_l = u_{l+N}$, i.e.

$$e^{-ikla} = e^{-ik(l+N)a} \quad \text{or} \quad e^{-ikNa} = 1.$$

Hence, $kNa = 2n\pi$, where n is any integer, or

$$k = 0, \pm\frac{2\pi}{aN}, \pm\frac{2\pi}{a}\frac{2}{N}, \ldots, \pm\frac{2\pi\frac{1}{2}N}{a}\frac{1}{N}, \tag{9}$$

i.e. there are approximately N possible values of k, evenly distributed throughout k-space with an interval $2\pi/Na$. Thus the density of states in k-space, $W(k)$—that is, the number of possible modes of vibration per unit interval in k—is a constant given by $2\pi W(k)/Na = 1$ or $W(k) = Na/2\pi$.

Of more interest, however, is the density of states as a function of frequency, $D(\omega)$. Now the number of states in a small frequency interval $d\omega$ will be

$$D(\omega)\,d\omega = W(k)\,dk = W(k)\frac{dk}{d\omega}\,d\omega; \tag{10}$$

therefore

$$D(\omega) = \frac{W(k)}{d\omega/dk}. \tag{11}$$

For the one-dimensional linear chain with nearest-neighbour interactions, it is easy to show that equation (11) becomes

$$D(\omega) = \frac{2N}{\pi(\omega_0^2 - \omega^2)^{\frac{1}{2}}}, \qquad \omega_0 = \sqrt{\frac{4\gamma}{m}}. \tag{12}$$

The factor 2 arises because ω is always positive, so the negative and positive regions of k-space both contribute to the range $d\omega$.

2.3 Extension to all-neighbour forces

The simple form of dispersion curve shown in Figure 2 is a result of assuming nearest-neighbour forces only. Extension to further neighbours is quite simple in one dimension. For instance, if $\gamma_1 \neq 0$ and $\gamma_2 \neq 0$, equation (5) gives

$$\omega^2 = \frac{2}{m}(\gamma_1 + \gamma_2 - \gamma_1 \cos ka - \gamma_2 \cos 2ka). \tag{13}$$

Figure 3 illustrates the case $\gamma_1 = \gamma_2$.

2.4 More than one atom per unit cell

Considerable complication occurs when the atoms in the chain are not all equivalent. Variations may occur in the masses, the force constants, and the atomic spacing. However, the essential features can be illustrated if we

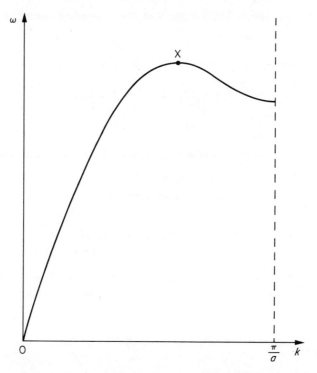

Figure 3. Phonon dispersion curve (first Brillouin zone, $k > 0$) for a linear chain of atoms with equal force constants for nearest- and second-nearest-neighbour interactions. X labels the point where the phonon group velocity is zero

assume even atomic spacing, a single, nearest-neighbour force constant γ, but different atomic masses, as shown in Figure 4.

The lattice spacing b is now the distance between like atoms (i.e. $b = 2a$), since the *unit cell* now contains two atoms. Thus we have two equations of motion, for the two types of atom (note that the suffix on u labels atom

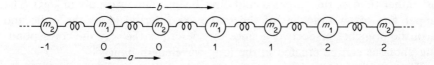

Figure 4. Linear chain of atoms of two species, having different masses but identical spacing and force constants. Although the distance apart is still a, the unit cell is of size $b = 2a$.

position, whereas the suffix on m and the bracketed number label the species of atom):

$$m_1 \frac{\partial^2 u_0(1)}{\partial t^2} = \gamma\{u_0(2) - u_0(1)\} + \gamma\{u_{-1}(2) - u_0(1)\}$$

$$= \gamma\{u_0(2) + u_{-1}(2) - 2u_0(1)\}; \qquad (14)$$

$$m_2 \frac{\partial^2 u_0(2)}{\partial t^2} = \gamma\{u_1(1) + u_0(1) - 2u_0(2)\}. \qquad (15)$$

We now use, in equations (14) and (15), the plane wave solution

$$u_l(1) = A e^{i(\omega t - klb)},$$
$$u_l(2) = B e^{i[\omega t - k(l+\frac{1}{2})b]}. \qquad (16)$$

This gives

$$-m_1 \omega^2 A = -2\gamma\{A - B \cos \tfrac{1}{2}kb\};$$
$$-m_2 \omega^2 B = -2\gamma\{B - A \cos \tfrac{1}{2}kb\}. \qquad (17)$$

The condition for equation (17) to have non-trivial solutions for A and B is that the determinant of coefficients of A and B must vanish, i.e.

$$\begin{vmatrix} 2\gamma - m_1 \omega^2 & -2\gamma \cos \tfrac{1}{2}kb \\ -2\gamma \cos \tfrac{1}{2}kb & 2\gamma - m_2 \omega^2 \end{vmatrix} = 0, \qquad (18)$$

giving

$$\omega^2 = \gamma\left(\frac{1}{m_1} + \frac{1}{m_2}\right) \pm \gamma\left[\left(\frac{1}{m_1} + \frac{1}{m_2}\right)^2 - \frac{4 \sin^2 \tfrac{1}{2}kb}{m_1 m_2}\right]^{\frac{1}{2}}. \qquad (19)$$

We can write this as

$$\omega^2 = \tfrac{1}{2}\omega_0^2[1 \pm \sqrt{1 - C \sin^2 \tfrac{1}{2}kb}] \qquad (20)$$

where $\omega_0^2 = 2\gamma(1/m_1 + 1/m_2)$ and C is commonly called a coupling coefficient, defined as

$$C = \frac{4m_1 m_2}{(m_1 + m_2)^2} = \frac{4\rho}{(1+\rho)^2}, \qquad \rho = \frac{m_1}{m_2}. \qquad (21)$$

The dispersion curve is now as shown in Figure 5 (note that had we plotted ω^2 rather than ω, the curves would have been symmetrical about $\tfrac{1}{2}\omega_0^2$). The upper and lower branches of the curve are referred to as the optic and acoustic branches respectively, since this describes the way the corresponding phonons can be created in the long-wavelength limit.

We would expect that as $m_1 \to m_2$, the above case should go over smoothly to the single atom case, and a quick check of equation (19) above will confirm this, but the dispersion curve would appear to be quite different. However, the anomaly is resolved if we remember that $b = 2a$, and when

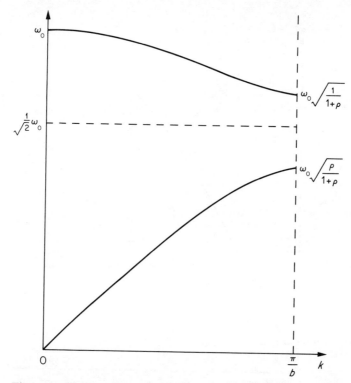

Figure 5. Phonon dispersion curve for a diatomic linear chain, with atoms of different masses m_1 and m_2 (where $\rho = m_1/m_2$, m_1 being the smaller mass) but identical spacing and force constants (between nearest neighbours only)

$m_1 \neq m_2$, the Brillouin zone is halved in size. Thus for the limiting case $m_1 = m_2$ the upper curve has been artificially 'folded back' from that part of the Brillouin zone where $\pi/2a < k \leqslant \pi/a$, as shown in Figure 6.

3. THREE-DIMENSIONAL CRYSTAL LATTICE

3.1 Normal mode frequencies in three dimensions

The biggest problem in generalizing the previous treatment to three dimensions is that the force constants become considerably more complicated, because the force on each atom due to displacement of its neighbours is a vector. Consider first how equation (1) for the force in one dimension may be rewritten:

$$F_0 = \sum_{l \neq 0} \gamma_l (u_l - u_0) = \sum_{l \neq 0} \gamma_l u_l - \sum_{l \neq 0} \gamma_l u_0. \tag{22}$$

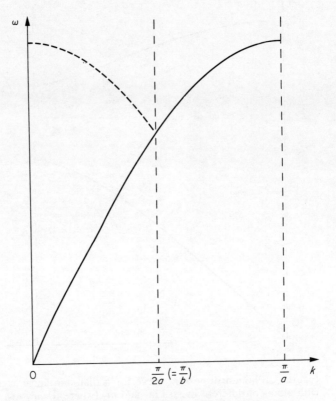

Figure 6. The true phonon dispersion curve for a monatomic linear chain (solid line) and when treated as the limiting case $m_1 = m_2$ for a diatomic linear chain (broken line), in which case the Brillouin zone is artifically reduced in size

If we define a special quantity γ_0 as $\gamma_0 = -\sum_{l \neq 0} \gamma_l$, F_0 reduces to

$$F_0 = \sum_{l=-\infty}^{+\infty} \gamma_l u_l. \tag{23}$$

The physical significance of γ_0 is simply that it represents the restoring force constant on the atom at the origin due to a displacement of itself.

In three dimensions, equation (23) can be generalized to

$$\mathbf{F}_0 = -\sum_l \mathbf{\Phi}_l \cdot \mathbf{u}_l, \tag{24}$$

(the minus sign is introduced simply for convenience, because $\mathbf{\Phi}$ then represents the second derivative of the equilibrium crystal potential energy) and the problem, which must be solved for the particular type of crystal structure under consideration, is to relate the tensor $\boldsymbol{\Phi}$ to the various force constants between neighbouring atoms.

If we use α and β as coefficients to represent the Cartesian components of the vectors, the force equation (24) can be written

$$F_{0,\alpha} = -\sum_l \sum_{\beta=1}^{3} \Phi_{l,\alpha\beta} u_{l,\beta}, \qquad \alpha = 1, 2, 3. \tag{25}$$

There are now three equations of motion for each atom, corresponding to the three degrees of freedom:

$$m \frac{\partial^2}{\partial t^2} u_{0,\alpha} = -\sum_l \sum_{\beta=1}^{3} \Phi_{l,\alpha\beta} u_{l,\beta} \tag{26}$$

and the three-dimensional form of a plane wave is

$$u_{l,\alpha}(\mathbf{k}) = A_\alpha \exp[i(\omega t - \mathbf{k} \cdot \mathbf{r}_l)], \tag{27}$$

for a wave travelling in the direction of $\mathbf{k}$, where $\mathbf{r}_l$ is the position vector of the lth atom. The precise form of $\mathbf{r}_l$ will depend on the crystal structure.

To facilitate the generalization to a unit cell containing more than one atom, it is convenient at this point to make two new definitions:

$$B_\alpha = m^{\frac{1}{2}} A_\alpha, \qquad \mathscr{D}_{\alpha\beta} = \frac{1}{m} \sum_l \Phi_{l,\alpha\beta} \exp(-i\mathbf{k} \cdot \mathbf{r}_l). \tag{28}$$

Using these quantities, substitution of equation (27) into the equations of motion (26) gives

$$\omega^2 B_\alpha = \sum_{\beta=1}^{3} \mathscr{D}_{\alpha\beta} B_\beta, \qquad \alpha = 1, 2, 3, \tag{29}$$

or in matrix form:

$$\omega^2 \mathbf{B} = \mathscr{D} \cdot \mathbf{B}. \tag{30}$$

This is a typical eigenvalue problem, and the solutions for ω^2 are the roots of the determinantal equation

$$\text{Det} (\mathscr{D} - \omega^2 \mathbf{I}) = 0. \tag{31}$$

The matrix $\mathscr{D}$ is commonly called the *dynamical matrix*, and the determinant is called the *secular determinant*.

It is interesting to note that, as might be expected, for the case of $\mathbf{k}$ in a symmetry direction the solutions (i.e. the normal modes of vibration) correspond to one longitudinal wave and two transverse waves (often degenerate). For non-symmetry directions, however, the normal modes do not necessarily have such a simple form.

3.2 Calculation of dynamical matrix for a BCC lattice

To proceed further it is necessary to specify the type of lattice and force constants to be included.

We shall consider the case of a body-centred cubic lattice which can be described by

$$\mathbf{r}_l = a\mathbf{l}, \qquad \mathbf{l} = (l_1, l_2, l_3), \tag{32}$$

where l_1, l_2, and l_3 are either all odd or all even integers.

We shall assume for simplicity that the interatomic force constants are purely radial, and that only nearest- and second-nearest-neighbour interactions are significant, with force constants γ and $R\gamma$, so that R represents the ratio of second-nearest to nearest force constants. We need to calculate the matrices Φ_l for each nearest and second-nearest neighbour ($l = 1\text{--}14$), and for $l = 0$.

If the atom 0 lies at the centre of a cube of side $2a$, then the eight nearest neighbours lie on the corners of the cube, along the [111], type directions, as shown in Figure 7. Consider atom 1, lying on the [111]-direction. Referring back to equation (25), we see that $-\Phi_{l,\alpha\beta}$ is the force on atom 0 in the α-direction resulting from unit displacement of atom l in the β-direction.

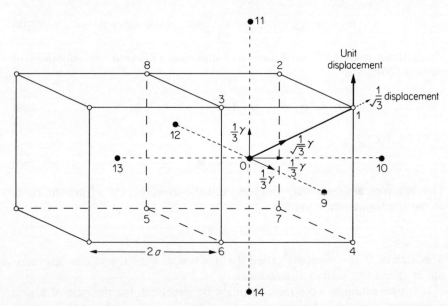

Figure 7. The BCC lattice, showing two of the unit cells, and all the nearest and second-nearest neighbours of the atom at the centre of the right-hand cell. This also represents the two-atom CsCl structure, where the two species are then represented by the solid and open circles respectively

Let us consider as an example the unit displacement of atom 1 in the x-, y- or z-direction. In each case this results in a radial component of displacement of $1/\sqrt{3}$, and since only radial forces are being considered, the resultant force on atom 0 is $(1/\sqrt{3})\gamma$. If we in turn resolve this force into its three Cartesian components, these can be written as $F_{0,x} = F_{0,y} = F_{0,z} = \frac{1}{3}\gamma$, Thus each Cartesian component of force due to unit displacement along any Cartesian axis is $\frac{1}{3}\gamma$, and this means that $-\Phi_{1,\alpha\beta} = \frac{1}{3}\gamma$ for all α and β. Thus

$$-\Phi_1 = \tfrac{1}{3}\gamma \begin{pmatrix} 1 & 1 & 1 \\ 1 & 1 & 1 \\ 1 & 1 & 1 \end{pmatrix}. \tag{33}$$

Exactly the same will be true for atom 5, in the $[\bar{1}\bar{1}\bar{1}]$-direction, i.e.

$$-\Phi_5 = -\Phi_1. \tag{34}$$

For the remaining atoms, along $[\bar{1}11]$-, $[1\bar{1}1]$-, and $[11\bar{1}]$-directions and their opposite neighbours, the matrices are respectively

$$-\Phi_2 = -\Phi_6 = \tfrac{1}{3}\gamma \begin{pmatrix} 1 & -1 & -1 \\ -1 & 1 & 1 \\ -1 & 1 & 1 \end{pmatrix}, \qquad -\Phi_3 = -\Phi_7 = \tfrac{1}{3}\gamma \begin{pmatrix} 1 & -1 & 1 \\ -1 & 1 & -1 \\ 1 & -1 & 1 \end{pmatrix},$$

$$-\Phi_4 = -\Phi_8 = \tfrac{1}{3}\gamma \begin{pmatrix} 1 & 1 & -1 \\ 1 & 1 & -1 \\ -1 & -1 & 1 \end{pmatrix}. \tag{35}$$

The second-nearest neighbours lie on the six faces of a cube of side $4a$. Thus for atom 9, lying along the $[100]$-direction, a unit displacement in the x-direction gives a force $R\gamma$ in that direction on the atom at 0, and orthogonal displacements have no effect. Thus the only non-zero matrix elements are

$$-\Phi_{9,xx} = -\Phi_{12,xx} = -\Phi_{10,yy} = -\Phi_{13,yy} = -\Phi_{11,zz} = -\Phi_{14,zz} = R\gamma. \tag{36}$$

Finally we must calculate Φ_0. Now this is simply the force on atom 0 due to unit displacement of itself. Thus a unit displacement of 0 along $0x$ is equivalent to a displacement of all the neighbours along $-0x$. Adding the components of force due to all the other atoms, we find

$$-\Phi_{0,xx} = -\tfrac{8}{3}\gamma - 2R\gamma, \qquad -\Phi_{0,xy} = 0, \qquad -\Phi_{0,xz} = 0, \text{ etc.} \tag{37}$$

Thus

$$-\Phi_0 = -2\gamma(\tfrac{4}{3} + R) \begin{pmatrix} 1 & 0 & 0 \\ 0 & 1 & 0 \\ 0 & 0 & 1 \end{pmatrix}. \tag{38}$$

We are now in a position to calculate $\mathscr{D}$ from equation (28), i.e.

$$\mathscr{D}_{\alpha\beta} = \sum_l \frac{1}{m} \Phi_{l,\alpha\beta} \exp(-i\mathbf{k}\cdot\mathbf{r}_l). \tag{39}$$

Thus, for example, the contributions to $m\mathscr{D}_{xx}$ are, using Figure 7, from atoms 1 and 5:

$$-\tfrac{1}{3}\gamma[\exp\{-i(k_xa+k_ya+k_za)\}+\exp\{i(k_xa+k_ya+k_za)\}],$$

from atoms 2 and 6:

$$-\tfrac{1}{3}\gamma[\exp\{-i(-k_xa+k_ya+k_za)\}+\exp\{i(-k_xa+k_ya+k_za)\}],$$

from atoms 3 and 7:

$$-\tfrac{1}{3}\gamma[\exp\{-i(k_xa-k_ya+k_za)\}+\exp\{i(k_xa-k_ya+k_za)\}],$$

from atoms 4 and 8:

$$-\tfrac{1}{3}\gamma[\exp\{-i(k_xa+k_ya-k_za)\}+\exp\{i(k_xa+k_ya-k_za)\}],$$

from atoms 9 and 12:

$$-R\gamma[\exp(-ik_x\cdot 2a)+\exp(ik_x\cdot 2a)],$$

from atom 0:

$$+2\gamma(\tfrac{4}{3}+R).$$

Hence

$$\mathscr{D}_{xx} = \frac{2\gamma}{m}(-\tfrac{4}{3}\cos k_xa\,\cos k_ya\,\cos k_za - R\cos 2k_xa + \tfrac{4}{3} + R). \tag{40}$$

Other elements of $\mathscr{D}$ can be just as readily obtained, for example,

$$\mathscr{D}_{xy} = \mathscr{D}_{yx} = +\tfrac{8}{3}\gamma\sin k_xa\,\sin k_ya\,\cos k_za. \tag{41}$$

Using definitions such as

$$c_x = \cos k_xa, \qquad s_x = \sin k_xa, \qquad c_{2x} = \cos 2k_xa,$$
$$\delta = 1 + \tfrac{3}{4}R - c_xc_yc_z, \tag{42}$$

the elements of $\mathscr{D}$ can be considerably simplified. The final result is

$$\mathscr{D}(\mathbf{k}) = \frac{8\gamma}{3m}\begin{pmatrix} \delta - \tfrac{3}{4}Rc_{2x} & s_xs_yc_z & s_xc_ys_z \\ s_xs_yc_z & \delta - \tfrac{3}{4}Rc_{2y} & c_xs_ys_z \\ s_xc_ys_z & c_xs_ys_z & \delta - \tfrac{3}{4}Rc_{2z} \end{pmatrix}. \tag{43}$$

3.3 Two-atom unit cell

The principal difference in the equations, in the case of more than one atom per unit cell, is that we have three equations of motion for each of the atoms. If we allow l to label all atoms, of whichever sort (note this is the alternative procedure to the one adopted in section 2.4), then the force constant matrices will have the same form as before, but it will be necessary to label Φ and $\mathscr{D}$, and of course the masses m and the amplitudes $\mathbf{B}$, with the type of atom.

The reason for introducing $\mathbf{B}$ and $\mathscr{D}$ originally is that they must now take the form:

$$\mathbf{B}(\kappa) = m_\kappa^{\frac{1}{2}} \mathbf{A}(\kappa), \qquad \mathscr{D}(\kappa, \kappa') = (m_\kappa m_{\kappa'})^{-\frac{1}{2}} \sum_l \Phi_l(\kappa, \kappa') \exp(-i\mathbf{k} \cdot \mathbf{r}_l), \quad (44)$$

where κ and κ' label the n different atoms in the unit cell. The equations of motion are

$$\omega^2 \mathbf{B}(\kappa) = \mathscr{D}(\kappa, \kappa) \cdot \mathbf{B}(\kappa) + \sum_{\kappa' \neq \kappa} \mathscr{D}(\kappa, \kappa') \cdot \mathbf{B}(\kappa'). \quad (45)$$

Equations (45) can be written more compactly as

$$\omega^2 \mathscr{B} = \mathscr{D}' \cdot \mathscr{B}, \quad (46)$$

where $\mathscr{D}'$ is a $3n \times 3n$ matrix and $\mathscr{B}$ is a column vector whose elements are $B_x(1)$, $B_y(1)$, $B_z(1)$, $B_x(2)$, etc.

Again we need to specify the type of lattice before proceeding any further. One of the simplest types of crystal structure having two atoms per unit cell is the CsCl structure, whose atoms form a BCC lattice having one type of atom at the cube centre and the other type on the corners of the cube, as shown in Figure 7. We shall assume as before that all the forces are radial and also that the force constants are the same for the two species of atom, so that we need consider only the difference in masses for the two types of atom. (In practice of course it would be physically unrealistic to consider only two force constants for an ionic crystal such as CsCl where the forces are quite long ranged.)

The matrices Φ will have exactly the same form as before, except that now they must also be labelled with the species of atom, and the first eight will couple different types whereas the next six couple identical atoms. Thus some matrices will now be zero, as follows: $\Phi_l(1, 1) = \Phi_l(2, 2) = 0$, $l = 1\text{--}8$; $\Phi_l(1, 2) = \Phi_l(2, 1) = 0$, $l = 9\text{--}14$; and of course $\Phi_0(1, 2) = \Phi_0(2, 1) = 0$ by definition, and $\Phi_0(1, 1) = \Phi_0(2, 2) = (\frac{8}{3}\gamma + 2\gamma R)\mathbf{I}$, as before.

If we remember that the ratio of masses is $\rho = m_1/m_2$, and use the notation of (42), we obtain for $\mathscr{D}'$

$$\mathscr{D}'(\mathbf{k}) = \frac{8}{3}\gamma(m_1 m_2)^{-\frac{1}{2}} \begin{pmatrix} \rho^{-\frac{1}{2}}(1+r_x) & 0 & 0 & -c_x c_y c_z & s_x s_y c_z & s_x c_y s_z \\ 0 & \rho^{-\frac{1}{2}}(1+r_y) & 0 & s_x s_y c_z & -c_x c_y c_z & c_x s_y s_z \\ 0 & 0 & \rho^{-\frac{1}{2}}(1+r_z) & s_x c_y s_z & c_x s_y s_z & -c_x c_y c_z \\ -c_x c_y c_z & s_x s_y c_z & s_x c_y s_z & \rho^{\frac{1}{2}}(1+r_x) & 0 & 0 \\ s_x s_y c_z & -c_x c_y c_z & c_x s_y s_z & 0 & \rho^{\frac{1}{2}}(1+r_y) & 0 \\ s_x c_y s_z & c_x s_y s_z & -c_x c_y c_z & 0 & 0 & \rho^{\frac{1}{2}}(1+r_z) \end{pmatrix},$$

$$(47)$$

where the r's arise from the second-nearest-neighbour forces, and are defined by $r_x = \frac{3}{4}R(1 - c_{2x})$, etc.

3.4 Density of states

In order to make an analytical calculation of the density of states in three dimensions, it is necessary to integrate $(\nabla_k \omega)^{-1}$ (the three-dimensional equivalent of equation (11)) over a constant frequency surface in k-space. Since this is not normally feasible, the usual approach is to compute the density of states numerically by the method first described by Walker.[1] This is usually referred to as the 'root sampling method', since it builds up the density of states by finding the roots of the secular equation at a large number of points in the Brillouin zone.

Just as in one dimension, the allowed values of k are evenly distributed in k-space, with a density such that there are N states in the first Brillouin zone. The simplest approach is to solve the secular determinant over a cubic mesh of points in k-space, and to plot a histogram of the computed frequencies. In practice, there are important refinements which involve interpolation between the calculated points,[2,3] but the principle remains the same.

In one dimension the edge of the first Brillouin zone was simply the point at which the wavelength was twice the lattice spacing and all higher values of k could be reflected back into the first Brillouin zone by simply subtracting a *reciprocal lattice vector*, $2\pi/a$.

In three dimensions, we need to be concerned with planes of atoms. (As a matter of fact the dispersion curve calculations of section 2.1 will also hold for vibrations of planes of atoms for waves in high symmetry directions, provided that the γ_l are reinterpreted as the interplanar force constants.)

For the two-atom unit cell of the CsCl structure, whose space lattice is simple cubic, the Brillouin zone is also a cube (but with its faces a perpendicular distance $\pi/2a$ from the origin since the unit cell has a side of $2a$). For a one-atom BCC lattice, on the other hand, the most widely spaced planes of atoms are the (110) planes, with spacing $\sqrt{2}a$, so that the limiting wavelength is $2\sqrt{2}a$, and (remembering that the wave vector is defined as $|k| = 2\pi/\lambda$) the first Brillouin zone is that volume bounded by the 12 (110)-type planes, each a perpendicular distance $\pi/\sqrt{2}a$ from the origin.

It is not necessary to calculate frequencies over the whole Brillouin zone, of course. Just as in one dimension, we need only consider positive values of k_x, k_y, and k_z, which reduces calculations to the first octant. Moreover, symmetry reduces the calculations necessary still further. For a cubic lattice the x-, y-, and z-directions must be equivalent, so only one-third of the octant need be considered, and if we also remember that the (110) planes have mirror symmetry, this divides the unique portion of the Brillouin zone in half again, to only 1/48 of the whole zone (e.g. that part lying between the [100]-, [110]-, and [111]-directions).

It is worth noting at this point that for the BCC lattice the [111] dispersion curves do not repeat outside the Brillouin zone boundary, but only after twice this distance. The reason is that the (111) planes are closely spaced, and can support waves of very short wavelength (i.e. large wave vector). However, these do not lie within the first Brillouin zone because they can be reflected back into the zone (in fact, on to the zone boundary on the line joining the [$\bar{1}\bar{1}$1]- and [001]-directions) by subtracting a [110] reciprocal lattice vector.

3.5 Van Hove critical points

If we refer back to the analytical expression for the density of states in one dimension (equation (12)) we can see that there is a singularity at $\omega = \omega_0$. The reason for this is easy to understand if we re-examine Figure 2. The density of points in k-space is uniform, so that at any point on the curve where the slope approaches zero there will be a very large density of points in ω-space. For the simple model illustrated in Figure 2 this occurs at the Brillouin zone boundary, where $\omega = \omega_0$. However, any point on a dispersion curve where the slope (i.e. the group velocity of the phonons) is zero will create a singularity in the density of states—for instance the point X in Figure 3—and these are known as *Van Hove singularities*.

In three dimensions a similar feature occurs wherever $\nabla_k(\omega) = 0$, except that whereas in equation (12) the effect was proportional to $(|\omega^2 - \omega_0^2|)^{-\frac{1}{2}}$, in three dimensions the corresponding contribution is proportional to the less drastic $(|\omega^2 - \omega_0^2|)^{\frac{1}{2}}$. The latter frequency dependence does not create a singularity at ω_0, merely a cusp or critical point in the curve as the extra contribution starts or ceases. Nevertheless, one of the most characteristic features of a three dimensional density of states curve is the series of Van Hove critical points, or discontinuities in the slope of the curve, and in many cases these can be associated with points of zero slope on the phonon dispersion curves in high symmetry directions.

4. COMPUTER PROGRAM

4.1 Computation of the density of states

In a normal experimental determination of the density of states, the actual measurements are made on the dispersion curves in symmetry directions in the crystal. The commonest method is probably inelastic neutron scattering, although the pioneering work by Walker[1] was based on X-ray diffuse scattering measurements.

The force constants (typically five to nine constants are determined, including some non-radial forces) are then found by solving the secular determinant along the symmetry directions and varying the constants to obtain the best fit to the experimental data.

For the purposes of the simulation program to be described here, this approach is reversed, and the user sets the ratio of the two force constants being considered. This value is then used to generate a theoretical set of dispersion curves so that the effect of the two constants on the dispersion curves can be examined.

The final determination of the density of states follows the normal experimental approach, using the values chosen for the force constants in the secular determinant and evaluating the determinant over a mesh of points in **k**-space to build up a histogram of the density of states.

4.2 Program details

Since numerical values are of less interest than the shape of the various curves, the program output consists entirely of curves, with arbitrary scales on the axes, plotted on the line printer (although it would be quite a simple task for the reader to alter the printing subroutines to have the output generated on a graph plotter). This means that it is unnecessary to specify values for the lattice spacing, force constant, or atomic mass, as it is only the ratios of the force constants and masses that affect the shape of the curves.

The program is based on the two subroutines D1 and D2, which set up the dynamical matrices for the one-atom and two-atom cases respectively. Each subroutine then calls a standard library subroutine to evaluate the secular determinant, whose eigenvalues are the squares of the frequencies of the normal modes of vibration. The program has been developed using the NAG subroutine F02AAF, which solves the simple eigenvalue problem $\mathbf{A} \cdot \mathbf{x} = \lambda \mathbf{x}$ for a real symmetric matrix. Any similar subroutine may be used instead, but it should be noted that the NAG subroutine returns the eigenvalues in ascending order. This fact is utilized in the main program, so if the subroutine chosen does not do this, it will be necessary to sort the values immediately after the subroutine call.

After reading in the necessary data (described in the next subsection) the program calculates the frequencies of the three normal modes at equal intervals of k, along the three symmetry directions [100], [110], and [111]. The subroutine PRDC then outputs the dispersion curves, in graphical form, on the line printer. The ω- and k-axes both have annotated scales with arbitrary units, but the respective scales are the same for all three sets of dispersion curves. Where the curves, or individual points on them, are degenerate, they are denoted by stars instead of crosses. For the one-atom case, the [111] dispersion curves do not repeat outside the Brillouin zone (as

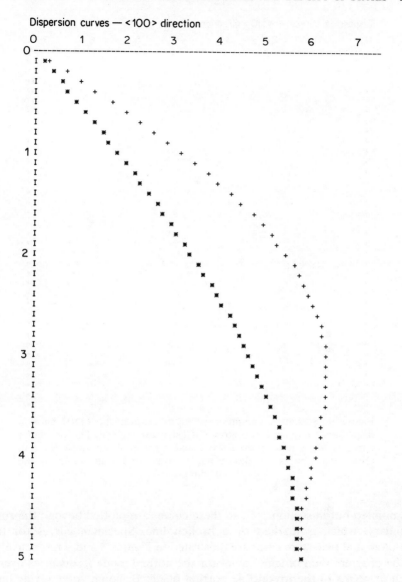

Figure 8. Specimen computer program output: the [100] phonon dispersion curve for a one-atom BCC crystal lattice. This particular curve is for the force constant ratio $R = 0.8$. DCXSC and DCYSC have been specified as 25 and 75 respectively; these values are smaller than would normally be specified for a line printer, but would be suitable for a terminal

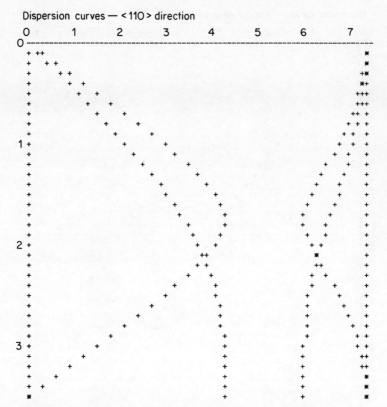

Figure 9. Specimen computer program output: the [110] phonon dispersion curve for a two-atom CsCl structure crystal. This particular curve is for a mass ratio $\rho = 0.5$ and force constant ratio $R = 0.0$. (Note that the latter choice has resulted in a soft mode in the [110]-directions)

was pointed out in section 3.4), so these curves are plotted beyond the zone boundary, which is marked by a broken line. Specimen curves for the one-atom and two-atom cases are illustrated in Figures 8 and 9 respectively.

The program then obtains values for the normal mode frequencies over a mesh of points in the irreducible portion of the Brillouin zone, giving only half weighting to points on the sides of the portion, since they are shared with the neighbouring portion. As the frequencies are calculated, they are immediately stored in histograms, which when completed form the densities of states for each of the modes of vibration. Instead of forming a simple histogram, each frequency value is made to contribute to the two frequency intervals centred above and below it, the contributions being weighted according to the closeness of the frequency value to the centre of the frequency interval. This weighting is a crude form of the refined calculations mentioned earlier, but it still results in a considerably smoother histogram for a given mesh size.

The subroutine PRDOS is used to plot the densities of states, again on the line printer. Three histograms, corresponding to the low-, medium-, and high-frequency normal modes, are output first. These do not necessarily correspond to any particular polarization, since the frequencies may cross over, and in any case the polarizations are not strictly longitudinal or transverse except in symmetry directions. For the two-atom case, acoustic and optic branches are combined in the same histogram since the frequencies do not overlap.

The total density of states is output last. The scales of the densities of states are again in arbitrary units, but the frequency scales do correspond to those of the dispersion curves, to facilitate comparison between them. Figure 10 illustrates typical output for the density of states, and a number of Van Hove critical points can clearly be seen.

4.3 Data for the program

Only six variables need to be specified as data. These are:

> DCXSC   in I3 format
> DCYSC   in I3 format

for each set of curves required:

> ATOMS   in I3 format
> INT   in I3 format
> R   in F6.2 format

for the two-atom case only:

> RO   in F6.2 format.

The function of these variables is as follows: DCXSC and DCYSC, which need to be specified only once, control the scale of the x- and y-axes of the output. The user may wish to experiment with these to obtain the most presentable size and relative scaling for the particular line printer (or terminal) in use. Once these are established, the values could be incorporated into the program.

ATOMS should be given the value 1 or 2 respectively to simulate the appropriate lattice—one-atom BCC or two-atom CsCl structure. The program will plot as many sets of curves as required and a value for ATOMS of zero must be specified to terminate the program.

INT sets the scale of the mesh for the density of states calculation. Obviously, the greater the number of mesh points the smoother the resulting histograms, but the number of points, and resulting computing time, will increase roughly as INT cubed. The user is left to set the size of the mesh

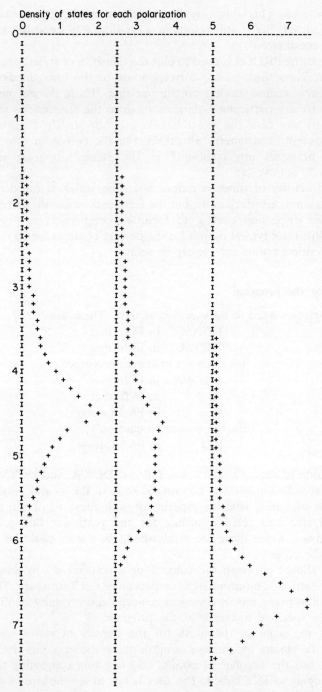

Figure 10. Specimen computer program output: the phonon densities of states. These particular curves are for a one-atom BCC lattice with $R = 0.8$ (a) Densities of states for the low-, medium-, and high-frequency polarizations; (b) total density of states

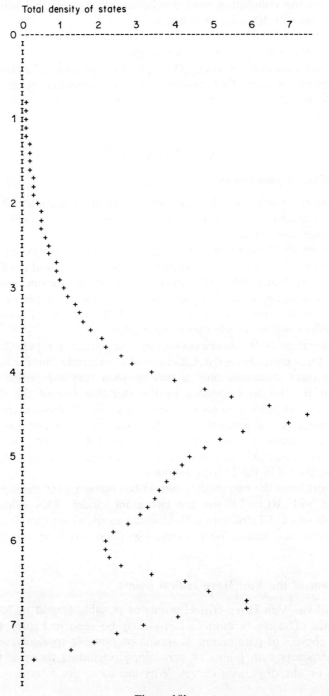

Figure 10b

according to the computing time available—a value of INT which gives reasonably smooth densities of states is 50.

R is the ratio of the second nearest to nearest-neighbour force constant, and any positive (or zero) value may be specified.

RO is only required for ATOMS = 2, and is the ratio of masses of the different species of atom. Only positive values are physically realistic, and to simulate all possibilities it is necessary only to investigate a range of values from 0 to 1.

5. EXERCISES FOR STUDENTS

5.1 Variation of parameters

The student can investigate the effect of varying the parameters R and RO for the two models. It is of particular interest to see the effects as the limiting values are approached.

In the limit as R becomes very large, the models approximate to two interpenetrating but non-interacting simple cubic lattices, and it will be seen that the second half of the [100] dispersion curve in the one-atom model becomes a reflection of the first half, because the larger simple cubic lattice has a Brillouin zone only half as long in the [100]-directions. Try and predict what the effect will be in the two-atom model.

Negative values of R (corresponding to a negative, or repulsive, second-neighbour force constant) are not disallowed in principle, but for the case of two radial force constants only it will be seen that the lattice becomes unstable at R = 0.0, as evidenced by the fact that one of the dispersion curves in the [110]-direction goes to zero. (A mode of vibration dropping to zero frequency is a characteristic of the onset of instability, and is often called a 'soft mode'.) This is because, without the stabilizing effect of (positive) second-neighbour forces, there is no resistance to shear of the (110) atom planes in the [$\bar{1}$10]-directions.

The results from the two models should be compared for the same values of R, and with RO = 1.0 for the two-atom model. This value of RO corresponds to a CsCl structure with identical atoms which therefore reduces to a one-atom BCC lattice. Why are the dispersion curves not identical?

5.2 Location of the Van Hove critical points

As many of the Van Hove critical points as possible should be located on the densities of states (a number can clearly be seen in Figure 10) for a variety of choices of parameters. It should be possible to associate most of the critical points with points of zero slope (including those at the zone boundary) on the dispersion curves. Why are there not a large number of critical points associated with zero slope positions elsewhere in the Brillouin zone?

5.3 Extension to face-centred cubic lattice

Readers with a reasonable programming ability should try to modify the program to perform the same calculations for a one-atom, face-centred cubic lattice (FCC). (The equivalent of a two-atom CsCl structure does not exist because there are four atoms to each cube, rather than two.) The only parts of the program which are affected are: the subroutine which calculates the dynamical matrix; the lengths of the dispersion curves; and also in consequence of the different Brillouin zone, the limits on the DO loops creating the mesh of points to calculate the densities of states.

The dynamical matrix for the FCC lattice is

$$\mathscr{D}(\mathbf{k}) = \frac{2\gamma}{m} \begin{pmatrix} \Delta + c_y c_z - R c_{2x} & s_x s_y & s_x s_z \\ s_x s_y & \Delta + c_x c_z - R c_{2y} & s_y s_z \\ s_x s_z & s_y s_z & \Delta + c_x c_y - R c_{2z} \end{pmatrix}, \quad (48)$$

where $\Delta = 2 + R - (c_x c_y + c_y c_z + c_z c_x)$.

The first Brillouin zone is slightly more complicated, being bounded partly by six (100)-type planes a perpendicular distance π/a from the centre, and partly by eight (111)-type planes a perpendicular distance $\sqrt{3}\pi/2a$ from the centre. The conditions for a mesh point to lie within the zone will therefore be $k_x + k_y + k_z \leq 3\pi/2a$, $k_x \leq \pi/a$ and $k_z \leq k_y \leq k_x$ (the last condition is imposed to restrict the mesh to the irreducible 1/48 of the Brillouin zone). The limits of the dispersion curves are hence $k_x = k_y = k_z = \pi/2a$ for the [111]-direction, $k_x = \pi/a$ for the [100]-direction, and $k_x = k_y = 3\pi/4a$ for the [110]-direction. However, in an analogous fashion to the [111] curves of the BCC lattice, the [110] curves do not repeat outside the zone boundary, but only after the point $k_x = k_y = \pi/a$.

REFERENCES

1. C. B. Walker, *Phys. Rev.*, **103**, 547 (1956).
2. G. Gilat and G. Dolling, *Phys. Lett.*, **8**, 304 (1964).
3. G. Gilat and L. J. Raubenheimer, *Phys. Rev.*, **144**, 390 (1966).

LATTICE DYNAMICS PROGRAM

```
      INTEGER ATOMS,INT,KX,KY,KZ,K,LIM,INT1,INT2,INT3,DIR,J,I,L,M,WINT,
     1.       IFAIL,DCYSC,DCXSC,DOSXSC,CH(120)
      REAL PI,R,WMAX,NWMAX,WT,WFRAC,QX,QY,QZ,W(3,6,200),NW(6,200),
     1    N(200),RO,NMAX
      COMMON CH
C
C     ATOMS MUST BE SET TO 1 FOR THE CASE OF THE SINGLE ATOM, B.C.C.
C     LATTICE AND 2 FOR THE CASE OF THE TWO ATOM CSCL LATTICE.
C     J IS THUS THE NUMBER OF BRANCHES OF THE DISPERSION CURVES.
C     R IS THE RATIO OF SECOND-NEAREST TO NEAREST NEIGHBOUR FORCE
C     CONSTANTS. IN THE LATTER CASE, RO IS THE RATIO OF ATOMIC MASSES.
C     R MUST BE .GE. ZERO FOR THE MODELS CHOSEN, AND RO MUST BE .GT.
C     ZERO. VALUES OF RO .LE. ONE WILL COVER ALL EVENTUALITIES.
C     INT IS AN INTEGER, THE NUMBER OF POINTS IN K-SPACE IN EACH
C     DIRECTION TO BE USED IN THE CALCULATIONS OF THE DENSITY OF
C     STATES.
C
      IFAIL=0
      PI=4.0*ATAN(1.0)
      READ(1,1001)DCXSC
C
C     DCXSC CONTROLS THE SCALE OF THE K AXIS OF THE DISPERSION
C     CURVES. 40 IS SUGGESTED AS A SUITABLE VALUE, BUT IN ANY CASE IT
C     SHOULD NOT EXCEED 57 UNLESS THE SIZE OF ARRAY W IS CHANGED.
C
      READ(1,1001)DCYSC
C
C     DCYSC CONTROLS THE SCALE OF THE W AXIS OF ALL THE CURVES
C     AND THE Y AXIS OF THE DENSITIY OF STATES CURVES.
C     117 IS SUGGESTED AS A SUITABLE VALUE FOR A 120 CHARACTER
C     LINE PRINTER, BUT IN ANY CASE IT MUST BE AT LEAST 3 LESS
C     THAN THE AVAILABLE NUMBER OF CHARACTERS, IT MUST BE
C     DIVISIBLE BY 3, AND IT SHOULD NOT EXCEED 120 UNLESS THE
C     DIMENSION OF CH AND THE FORMAT STATEMENTS WHERE IT IS
C     PRINTED ARE ALTERED ACCORDINGLY.
C
  100 READ(1,1001)ATOMS
 1001 FORMAT(I3)
      IF (ATOMS.EQ.0) GOTO 199
      READ(1,1001)INT
      READ(1,1002)R
 1002 FORMAT(F6.2)
      WRITE(2,1003)
 1003 FORMAT(/////)
      WRITE(2,1005)
 1005 FORMAT(48H DISPERSION CURVES AND DENSITIES OF STATES FOR A)
      IF (ATOMS.EQ.2) GOTO 101
      J=3
      WRITE(2,1006)
 1006 FORMAT(33H MONATOMIC B.C.C. CRYSTAL LATTICE)
  101 CONTINUE
      IF (ATOMS.EQ.1) GOTO 102
      J=6
      READ(1,1002)RO
      WRITE(2,1007)
 1007 FORMAT(35H DIATOMIC CSCL TYPE CRYSTAL LATTICE)
```

LATTICE DYNAMICS PROGRAM

```
  102 CONTINUE
      WRITE(2,1008)R
 1008 FORMAT(42H RATIO OF INTERATOMIC FORCE CONSTANTS = 1:,F8.4)
      IF (ATOMS.EQ.2) WRITE(2,1009)RO
 1009 FORMAT(28H RATIO OF ATOMIC MASSES = 1:,F8.4)
      WMAX=0.0
C
C     INTEGER COUNTERS K, KX, KY, KZ WILL BE USED TO CONTROL LOOPS OVER
C     MESHES IN K-SPACE. QX, QY, QZ WILL BE USED FOR THE ACTUAL VALUES
C     OF THE PRODUCT KA. THUS FOR THE ATOMS=1 CASE, THE MAXIMUM VALUE
C     OF EACH Q WILL BE PI ALONG THE <100> DIRECTIONS, AND PI/2 FOR THE
C     <110> AND <111> DIRECTIONS. HOWEVER, SINCE IN THE <111> DIRECTIONS
C     THE DISPERSION CURVES DO NOT REPEAT OUTSIDE THE BRILLOUIN ZONE
C     THESE ARE ALLOWED TO RUN TO THE REPEAT DISTANCE PI, AND THE EDGE
C     OF THE BRILLOUIN ZONE IS MARKED BY A BROKEN LINE ON THE PRINT-OUT.
C         FOR ATOMS=2 THE MAXIMUM VALUE OF EACH Q IS PI/2.
C
      INT1=DCXSC*2/ATOMS
      DO 103 K=1,INT1,1
      QX=FLOAT(K)*PI/FLOAT(INT1*ATOMS)
      QY=0.0
      QZ=0.0
      IF (ATOMS.EQ.1) CALL D1(1,K,QX,QY,QZ,R,W,IFAIL)
      IF (ATOMS.EQ.2) CALL D2(1,K,QX,QY,QZ,R,RO,W,IFAIL)
      IF (IFAIL.EQ.1) GOTO 199
      IF (WMAX.LT.W(1,J,K)) WMAX=W(1,J,K)
C
C     THE SUBROUTINE F02AAF IN THE SUBROUTINES D1 AND D2 RETURNS THE
C     EIGENVALUES OF W**2 IN ASCENDING ORDER, SO FINDING THE MAXIMUM
C     VALUE OF W(DIR,J,K) FOR ALL DIR AND K IS SUFFICIENT TO FIND THE
C     MAXIMUM FREQUENCY.
C
  103 CONTINUE
      INT2=IFIX(FLOAT(DCXSC)*SQRT(2.0))
      DO 104 K=1,INT2,1
      QX=FLOAT(K)*PI*0.5/FLOAT(INT2)
      QY=QX
      QZ=0.0
      IF (ATOMS.EQ.1) CALL D1(2,K,QX,QY,QZ,R,W,IFAIL)
      IF (ATOMS.EQ.2) CALL D2(2,K,QX,QY,QZ,R,RO,W,IFAIL)
      IF (IFAIL.EQ.1) GOTO 199
      IF (WMAX.LT.W(2,J,K)) WMAX=W(2,J,K)
  104 CONTINUE
      INT3=IFIX(FLOAT(DCXSC)*SQRT(3.0)*2.0/FLOAT(ATOMS))
      DO 105 K=1,INT3,1
      QX=FLOAT(K)*PI/FLOAT(INT3*ATOMS)
      QY=QX
      QZ=QX
      IF (ATOMS.EQ.1) CALL D1(3,K,QX,QY,QZ,R,W,IFAIL)
      IF (ATOMS.EQ.2) CALL D2(3,K,QX,QY,QZ,R,RO,W,IFAIL)
      IF (IFAIL.EQ.1) GOTO 199
      IF (WMAX.LT.W(3,J,K)) WMAX=W(3,J,K)
  105 CONTINUE
      WRITE(2,1010)
 1010 FORMAT(////37H  DISPERSION CURVES - <100> DIRECTION)
      CALL PRDC(INT1,1,J,DCYSC,W,WMAX)
```

LATTICE DYNAMICS PROGRAM


```
      WRITE(2,1011)
 1011 FORMAT(//37H  DISPERSION CURVES - <110> DIRECTION)
      CALL PRDC(INT2,2,J,DCYSC,W,WMAX)
      WRITE(2,1012)
 1012 FORMAT(//37H  DISPERSION CURVES - <111> DIRECTION)
      CALL PRDC(INT3,3,J,DCYSC,W,WMAX)
C
C     THE IRREDUCIBLE PORTION OF THE BRILLOUIN ZONE IS SPECIFIED BY
C     THE CONDITION KZ<KY<KX. THE BRILLOUIN ZONE BOUNDARY FOR ATOMS=2
C     (SIMPLE CUBIC) IS SIMPLY A CUBE OF SIDE PI/2. THUS ALLOWING VALUES
C     OF QX UP TO PI/2 TERMINATES ON THE BRILLOUIN ZONE BOUNDARY.
C         THE BRILLOUIN ZONE FOR ATOMS=1 (F.C.C. RECIPROCAL LATTICE) IS
C     A <110> PLANE, I.E. IT IS SPECIFIED BY QX+QY=PI. THUS ALLOWING
C     QX VALUES UP TO PI, AND SPECIFYING QY<QX AND ALSO QY<<PI-QX)
C     TERMINATES ON THE ZONE BOUNDARY. IT IS NECESSARY TO GIVE POINTS ON
C     THE ZONE BOUNDARY, OR THE BOUNDARIES OF THE IRREDUCIBLE PORTION,
C     REDUCED WEIGHTING OF ONE HALF. HOWEVER, IF KZ IS GIVEN HALF
C     INTEGER VALUES, TWO OF THE THREE SIDES, AND ALL THE EDGES OF THE
C     PORTION ARE AVOIDED, WHICH SIMPLIFIES THE POINTS OF REDUCED
C     WEIGHTING TO THOSE FOR QY=QX OR (PI-QX), AND QX=PI/2 (THE LATTER
C     ONLY MATERIALLY AFFECTS THE ATOMS=2 CASE).
C
      DO 106 WINT=1,200,1
        N(WINT)=0
        DO 106 I=1,6,1
          NW(I,WINT)=0
  106 CONTINUE
      DOSXSC=0
      INT2=INT
      IF (ATOMS.EQ.2) INT2=INT/2
      DO 107 KX=1,INT2,1
        QX=FLOAT(KX)*PI/FLOAT(INT)
        LIM=KX
        IF (KX.GT.INT/2) LIM=INT-KX
C
C     SETS LIM AS THE APPROPRIATE BOUNDARY FOR KY.
C
        DO 107 KY=1,LIM,1
          QY=FLOAT(KY)*PI/FLOAT(INT)
          DO 107 KZ=1,KY,1
            QZ=(FLOAT(KZ)-0.5)*PI/FLOAT(INT)
            IF (ATOMS.EQ.1) CALL D1(1,1,QX,QY,QZ,R,W,IFAIL)
            IF (ATOMS.EQ.2) CALL D2(1,1,QX,QY,QZ,R,RO,W,IFAIL)
            IF (IFAIL.EQ.1) GOTO 199
C
C     IT IS UNNECESSARY THIS TIME TO FILL ALL ELEMENTS OF THE ARRAY W
C     SINCE THE VALUES WILL IMMEDIATELY BE USED TO INCREMENT THE
C     APPROPRIATE ARRAY ELEMENTS OF THE DENSITY OF STATES.
C
            WT=1.0
            IF (KY.EQ.LIM.OR.KX.EQ.INT2) WT=0.5
            DO 107 I=1,J,1
              W(1,I,1)=FLOAT(DCYSC-1)*W(1,I,1)/WMAX
              WINT=IFIX(W(1,I,1))
              IF (DOSXSC.LT.WINT+2) DOSXSC=WINT+2
C
```

LATTICE DYNAMICS PROGRAM

```
C     THE SCALE OF THE X AXIS (WHICH IS W IN THIS CASE) IS MADE THE
C     SAME AS THE Y AXIS (ALSO W) OF THE DISPERSION CURVES.
C      THE SCALE LENGTH OF THE X AXIS IS SET TO THE LARGEST VALUE OF
C     FREQUENCY ANYWHERE IN THE BRILLOUIN ZONE IN CASE THERE ARE ANY
C     FREQUENCIES GREATER THAN WMAX, WHICH IS THE GREATEST FREQUENCY
C     IN ANY SYMMETRY DIRECTION.
C
                  WFRAC=W(1,I,1)-FLOAT(WINT)
                  NW(I,WINT+1)=NW(I,WINT+1)+(1.0-WFRAC)*WT
                  NW(I,WINT+2)=NW(I,WINT+2)+WFRAC*WT
C
C     TO GENERATE A SMOOTHER DENSITY OF STATES, RATHER THAN PLOT A
C     SIMPLE HISTOGRAM EACH CALCULATED FREQUENCY IS DIVIDED BETWEEN
C     THE TWO HISTOGRAM POINTS ABOVE AND BELOW, WEIGHTED ACCORDING TO
C     HOW CLOSE THE FREQUENCY IS TO EITHER POINT.
C
  107 CONTINUE
      NWMAX=0.0
      NMAX=0.0
      DO 108 WINT=1,200,1
        DO 109 I=1,3,1
          IF (ATOMS.EQ.2) NW(I,WINT)=NW(I,WINT)+NW(I+3,WINT)
          N(WINT)=N(WINT)+NW(I,WINT)
          IF (NWMAX.LT.NW(I,WINT)) NWMAX=NW(I,WINT)
  109   CONTINUE
        IF (NMAX.LT.N(WINT)) NMAX=N(WINT)
  108 CONTINUE
      CALL PRDOS(NW,NWMAX,N,NMAX,DCYSC,DOSXSC)
      GOTO 100
  199 STOP
      END




      SUBROUTINE D1(DIR,K,QX,QY,QZ,R,W,IFAIL)
      INTEGER DIR,K,I,J,IFAIL
      REAL QX,QY,QZ,R,DIAG,W(3,6,200),D(3,3),S(3),C(3),W2(3),
     1     WKSPC(3)
C
C     THIS SUBROUTINE SETS UP THE 3*3 DYNAMICAL MATRIX FOR THE ONE ATOM
C     UNIT CELL, AND USES A STANDARD LIBRARY SUBROUTINE F02AAF TO FIND
C     THE EIGENVALUES OF THE SECULAR DETERMINANT. ANY SIMILAR SUB-
C     ROUTINE WHICH FINDS THE EIGENVALUES OF A REAL SYMMETRIC MATRIX
C     MAY BE USED INSTEAD, AND THE EIGENVECTORS ARE NOT REQUIRED.
C
      C(1)=COS(QX)
      C(2)=COS(QY)
      C(3)=COS(QZ)
      S(1)=SIN(QX)
      S(2)=SIN(QY)
      S(3)=SIN(QZ)
      DIAG=1.0+0.75*R-C(1)*C(2)*C(3)
      D(1,1)=DIAG-0.75*R*COS(2.0*QX)
      D(2,2)=DIAG-0.75*R*COS(2.0*QY)
```

LATTICE DYNAMICS PROGRAM

```
      D(3,3)=DIAG-0.75*R*COS(2.0*QZ)
      D(1,2)=S(1)*S(2)*C(3)
      D(2,3)=C(1)*S(2)*S(3)
      D(3,1)=S(1)*C(2)*S(3)
      D(2,1)=D(1,2)
      D(3,2)=D(2,3)
      D(1,3)=D(3,1)
      CALL F02AAF(D,3,3,W2,WKSPC,0)
C
C     THE PARAMETERS OF F02AAF(D,N,N,W2,WKSPC,IFAIL) ARE AS FOLLOWS:
C     D IS THE DYNAMICAL MATRIX, OF DIMENSION NXN, W2 IS THE ARRAY
C     OF EIGENVALUES (THE SQUARES OF THE NORMAL MODE FREQUENCIES),
C     WKSPC IS AN ARRAY OF DIMENSION N PROVIDED FOR WORKING SPACE
C     AND IFAIL IS AN INTEGER DETERMINING THE MODE OF ACTION IN
C     THE EVENT OF FAILURE OF THE SUBROUTINE. THE VALUE 0 CAUSES
C     TERMINATION OF THE PROGRAM   WHENEVER AN ERROR IS DETECTED.
C
      DO 501 I=1,3,1
         IF(ABS(W2(I)).LT.0.00001) W2(I)=0.0
C
C     ROUNDING ERRORS MAY PRODUCE SMALL NEGATIVE VALUES IN
C     THE EIGENVALUES WHEN THE CORRECT VALUES ARE ZERO
C
         IF(W2(I).LT.0.0) GOTO 502
         W(DIR,I,K)=SQRT(W2(I))
  501 CONTINUE
      RETURN
  502 WRITE(2,5001)
 5001 FORMAT(42H SUBROUTINE D1 PRODUCES NEGATIVE FREQUENCY)
      IFAIL=1
      RETURN
      END


      SUBROUTINE D2(DIR,K,QX,QY,QZ,R,RO,W,IFAIL)
      INTEGER DIR,K,I,J,IFAIL
      REAL QX,QY,QZ,R,RO,ROOTRO,W(3,6,200),D(6,6),S(3),C(3),
     1     W2(6),WKSPC(6)
C
C     THIS SUBROUTINE IS SIMILAR TO D1, EXCEPT THAT IT SOLVES THE 6*6
C     MATRIX FOR THE TWO ATOM UNIT CELL.
C
      C(1)=COS(QX)
      C(2)=COS(QY)
      C(3)=COS(QZ)
      S(1)=SIN(QX)
      S(2)=SIN(QY)
      S(3)=SIN(QZ)
      ROOTRO=SQRT(RO)
      DO 801 I=1,6,1
         DO 801 J=1,6,1
            D(I,J)=0.0
  801 CONTINUE
```

LATTICE DYNAMICS PROGRAM

```
      D(1,1)=ROOTRO*(1.0+0.75*R-0.75*R*COS(2.0*QX))
      D(2,2)=ROOTRO*(1.0+0.75*R-0.75*R*COS(2.0*QY))
      D(3,3)=ROOTRO*(1.0+0.75*R-0.75*R*COS(2.0*QZ))
      D(4,4)=(1.0+0.75*R-0.75*R*COS(2.0*QX))/ROOTRO
      D(5,5)=(1.0+0.75*R-0.75*R*COS(2.0*QY))/ROOTRO
      D(6,6)=(1.0+0.75*R-0.75*R*COS(2.0*QZ))/ROOTRO
      DO 802 I=1,3,1
         D(I,I+3)=-C(1)*C(2)*C(3)
         D(I+3,I)=D(I,I+3)
  802 CONTINUE
      D(1,5)=S(1)*S(2)*C(3)
      D(1,6)=S(1)*C(2)*S(3)
      D(2,6)=C(1)*S(2)*S(3)
      D(5,1)=D(1,5)
      D(2,4)=D(1,5)
      D(4,2)=D(1,5)
      D(6,1)=D(1,6)
      D(3,4)=D(1,6)
      D(4,3)=D(1,6)
      D(6,2)=D(2,6)
      D(3,5)=D(2,6)
      D(5,3)=D(2,6)
      CALL F02AAF(D,6,6,W2,WKSPC,0)
      DO 803 I=1,6,1
         IF (ABS(W2(I)).LT.0.00001) W2(I)=0.0
C
C     ROUNDING ERRORS MAY PRODUCE SMALL NEGATIVE VALUES IN
C     THE EIGENVALUES WHEN THE CORRECT VALUES ARE ZERO
C
         IF (W2(I).LT.0.0) GOTO 804
         W(DIR,I,K)=SQRT(W2(I))
  803 CONTINUE
      RETURN
  804 WRITE(2,8001)
 8001 FORMAT(42H SUBROUTINE D2 PRODUCES NEGATIVE FREQUENCY)
      IFAIL=1
      RETURN
      END


      SUBROUTINE PRDC(INT,DIR,J,DCYSC,W,WMAX)
      INTEGER INT,DIR,J,I,K,L,M,KMAX,XAXIS,YAXIS,SPACE,CROSS,SUB,
     1        DCYSC,STAR,PT(6),CH(120),NOUGHT,ONE,TWO,THREE,
     2        FOUR,FIVE,SIX,SEVEN,EIGHT,NINE,DCYSC2,SCALE
      REAL WMAX,W(3,6,200)
      COMMON CH
      DATA XAXIS,YAXIS,SPACE,CROSS,STAR/1HI,1H-,1H ,1H+,1H*/
      DATA NOUGHT,ONE,TWO,THREE,FOUR/1H0,1H1,1H2,1H3,1H4/
      DATA FIVE,SIX,SEVEN,EIGHT,NINE/1H5,1H6,1H7,1H8,1H9/
C
C     THIS SUBROUTINE PRODUCES A GRAPH OF THE DISPERSION CURVES ON
C     THE LINEPRINTER.
C
```

LATTICE DYNAMICS PROGRAM

```
          DCYSC2=DCYSC+2
          DO 601 L=1,DCYSC2,1
601          CH(L)=SPACE
          CH(2)=NOUGHT
          CH(12)=ONE
          CH(22)=TWO
          CH(32)=THREE
          CH(42)=FOUR
          CH(52)=FIVE
          CH(62)=SIX
          CH(72)=SEVEN
          CH(82)=EIGHT
          CH(92)=NINE
          CH(102)=NOUGHT
          CH(112)=ONE
          WRITE(2,1601) (CH(K),K=1,DCYSC2)
          DO 602 L=2,DCYSC2,1
602          CH(L)=YAXIS
          CH(1)=NOUGHT
          WRITE(2,1601) (CH(K),K=1,DCYSC2)
1601     FORMAT(1H ,120A1)
          DO 603 I=1,INT,1
            DO 604 L=1,DCYSC2,1
              CH(L)=SPACE
              IF(DIR.EQ.3.AND.J.EQ.3.AND.I.EQ.INT/2
     1          .AND.L.EQ.(L/2)*2) CH(L)=YAXIS
604          CONTINUE
            CH(1)=XAXIS
            DO 605 L=1,J,1
              PT(L)=IFIX(W(DIR,L,I)*FLOAT(DCYSC-1)/WMAX+1.5)
              SUB=PT(L)
              IF (CH(SUB).EQ.CROSS.OR.CH(SUB).EQ.STAR) GOTO 606
              CH(SUB)=CROSS
              GOTO 605
606          CH(SUB)=STAR
605          CONTINUE
            SCALE=I/10
            IF(SCALE*10.EQ.I) GOTO 607
            WRITE(2,1602) (CH(K),K=1,DCYSC2)
1602     FORMAT(2H ,120A1)
            GOTO 603
607          IF(SCALE.LT.10) GOTO 608
            SCALE=SCALE-10
            GOTO 607
608          WRITE(2,1603)SCALE, (CH(K),K=1,DCYSC2)
1603     FORMAT(1H ,I1,120A1)
603     CONTINUE
          RETURN
          END
```

LATTICE DYNAMICS PROGRAM

```
      SUBROUTINE PRDOS( NW, NWMAX, N, NMAX, DCYSC, DOSXSC )
      INTEGER I, L, CH( 120 ), DCYSC, DOSXSC, XAXIS, YAXIS, SPACE, CROSS,
     1         INT2, SUB, NOUGHT, ONE, TWO, THREE, FOUR, FIVE,
     2         SIX, SEVEN, EIGHT, NINE, DCYSC2, SCALE, K
      REAL NW( 6, 200 ), NWMAX, N( 200 ), NMAX
      COMMON CH
      DATA XAXIS, YAXIS, SPACE, CROSS/1HI, 1H-, 1H , 1H+/
      DATA NOUGHT, ONE, TWO, THREE, FOUR/1H0, 1H1, 1H2, 1H3, 1H4/
      DATA FIVE, SIX, SEVEN, EIGHT, NINE/1H5, 1H6, 1H7, 1H8, 1H9/
C
C     THIS SUBROUTINE PRODUCES ON THE LINEPRINTER GRAPHS OF THE DENSITY
C     OF STATES FOR EACH POLARIZATION, AND THE TOTAL DENSITY OF STATES.
C
      WRITE( 2, 1200 )
 1200 FORMAT( //////41H  DENSITY OF STATES FOR EACH POLARIZATION)
      DCYSC2=DCYSC+2
      DO 201 L=1, DCYSC2, 1
 201     CH( L )=SPACE
      CH( 2 )=NOUGHT
      CH( 12 )=ONE
      CH( 22 )=TWO
      CH( 32 )=THREE
      CH( 42 )=FOUR
      CH( 52 )=FIVE
      CH( 62 )=SIX
      CH( 72 )=SEVEN
      CH( 82 )=EIGHT
      CH( 92 )=NINE
      CH( 102 )=NOUGHT
      CH( 112 )=ONE
      WRITE( 2, 1201 ) ( CH(K), K=1, DCYSC2 )
      DO 202 L=2, DCYSC2, 1
 202     CH( L )=YAXIS
      CH( 1 )=NOUGHT
      WRITE( 2, 1201 ) ( CH(K), K=1, DCYSC2 )
 1201 FORMAT( 1H , 120A1 )
      DOSXSC=DOSXSC+2
      DO 203 I=1, DOSXSC, 1
      DO 204 L=1, DCYSC2, 1
 204       CH( L )=SPACE
      INT2=DCYSC/3
      DO 205 L=1, 3, 1
          SUB=IFIX( NW( L, I+1 )*FLOAT( INT2 )/NWMAX+1.5 )+INT2*( L-1 )
          CH( SUB )=CROSS
 205  CONTINUE
      CH( 1 )=XAXIS
      CH( INT2+1 )=XAXIS
      INT2=2*INT2
      CH( INT2+1 )=XAXIS
      SCALE=I/10
      IF( SCALE*10.EQ.I )  GOTO 206
      WRITE( 2, 1202 ) ( CH(K), K=1, DCYSC2 )
 1202    FORMAT( 2H , 120A1 )
      GOTO 203
 206  IF( SCALE.LT.10 ) GOTO 207
      SCALE=SCALE-10
```

LATTICE DYNAMICS PROGRAM

```
        GOTO 206
207     WRITE(2,1203)SCALE, (CH(K),K=1,DCYSC2)
1203    FORMAT(1H ,I1,120A1)
203 CONTINUE
        WRITE(2,1300)
1300 FORMAT(/////25H   TOTAL DENSITY OF STATES)
        DO 301 L=1,DCYSC2,1
301     CH(L)=SPACE
        CH(2)=NOUGHT
        CH(12)=ONE
        CH(22)=TWO
        CH(32)=THREE
        CH(42)=FOUR
        CH(52)=FIVE
        CH(62)=SIX
        CH(72)=SEVEN
        CH(82)=EIGHT
        CH(92)=NINE
        CH(102)=NOUGHT
        CH(112)=ONE
        WRITE(2,1301) (CH(K),K=1,DCYSC2)
        DO 302 L=2,DCYSC2,1
302     CH(L)=YAXIS
        CH(1)=NOUGHT
        WRITE(2,1301) (CH(K),K=1,DCYSC2)
1301 FORMAT(1H ,120A1)
        DO 303 I=1,DOSXSC,1
        DO 304 L=1,DCYSC2,1
304     CH(L)=SPACE
        SUB=IFIX(N(I+1)*FLOAT(DCYSC)/NMAX+1.5)
        CH(SUB)=CROSS
        CH(1)=XAXIS
        SCALE=I/10
        IF(SCALE*10.EQ.I)  GOTO 305
        WRITE(2,1302) (CH(K),K=1,DCYSC2)
1302    FORMAT(2H ,120A1)
        GOTO 303
305     IF(SCALE.LT.10) GOTO 306
        SCALE=SCALE-10
        GOTO 305
306     WRITE(2,1303)SCALE, (CH(K),K=1,DCYSC2)
1303    FORMAT(1H ,I1,120A1)
303 CONTINUE
        RETURN
        END
```

Physics Programs
Edited by A. D. Boardman
© 1980 John Wiley & Sons Ltd.

CHAPTER 10

Electron Energy Bands in a One-dimensional Periodic Potential

R. D. CLARKE and D. J. MARTIN

1. INTRODUCTION

A particularly demanding area encountered in any course on solid-state physics is that of the energies of electrons in crystals.[1] Familiarity with this material is basic to a proper understanding of electrical phenomena in metals and semiconductors; perhaps the most significant difficulty that arises is that any realistic treatment necessitates the use of numerical methods. Commonly, however, only qualitative arguments or rather unrealistic models, such as the Kronig–Penney model,[2] are presented. The computer program presented here enables the user to investigate electron energies in a system where the form, magnitude, and period of the potential can be specified by the user. The results can be compared with those derived from approximate analytic treatments for certain ranges of the parameters. Exercises of this kind can supplement more conventional presentations and give students some familiarity with the basic methods employed in band structure calculations.

Any system involving particles will exhibit quantum-mechanical features if the de Broglie wavelength associated with the momentum of the particles is of the same order of magnitude or greater than a typical length over which the potential acting on the particles changes significantly. It is easy to show that this is the case for conduction electrons in a solid by the following considerations.

In the case of metals a typical free electron density (N) is $\sim 4 \times 10^{28} \, \text{m}^{-3}$. An ideal Fermi–Dirac gas of this density would have kinetic energy (E) per particle $\sim \hbar^2 (3\pi^2 N)^{\frac{2}{3}}/2m_e \sim 7 \times 10^{-19} \, \text{J}$, a velocity $(v) \sim \sqrt{2E/m_e} \sim 1.2 \times 10^6 \, \text{m s}^{-1}$, and a corresponding wavelength $\lambda = h/m_e v \sim 6 \times 10^{-10} \, \text{m}$. For a semiconductor $E \sim k_B T \sim 10^{-21} \, \text{J}$, so $v \sim 10^5 \, \text{m s}^{-1}$ and $\lambda \sim 7 \times 10^{-9} \, \text{m}$.

The potential acting on the electrons will vary significantly over the interatomic spacing $\sim 3 \times 10^{-10}$ m. It follows that quantum-mechanical methods are essential to tackle the problem.

The computer program presented here solves the Schrödinger equation to an accuracy of ~ 1 per cent for electrons in a one-dimensional potential, the form, magnitude, and period of which can be set by the user.

This study was limited to the one-dimensional case because it is designed as an aid to learning and deliberately avoids the numerical and conceptual complications of the three-dimensional case. It should, however, be noted that there are in fact some systems such as 'KCP' ($K_2Pt(CN)_4Br_{0.3} \cdot 3H_2O$) where the electron motion is in reality effectively confined to one dimension.[3]

One of the most significant factors influencing the behaviour of electrons in a typical metal or semiconductor is the fact that the potential to which they are subject is *periodic*. The reason for the periodicity is that the atoms are arranged in a regular crystal lattice. (Amorphous—randomized—metals and semiconductors exist but a rather different theoretical approach is then necessary.) The potential is periodic over many thousands of atoms, even if the material as a whole is polycrystalline.

2. ELECTRONS IN A PERIODIC POTENTIAL

2.1 Brillouin zones

It is not difficult to appreciate that the periodic potential due to the regular crystal lattice will affect the electron energies, and indeed to see in general terms what the effect will be. Electrons in free space have kinetic energy $E = \frac{1}{2}m_e v^2$. Knowing that for electrons in a solid quantum effects are important we can use the de Broglie relationship to express this as

$$E = \frac{h^2}{2m_e\lambda^2} \quad \text{or} \quad E = \frac{\hbar^2 k^2}{2m_e}$$

where $k(=2\pi/\lambda)$ is called the wave number. The corresponding wave function in free space is $\psi(x) = \exp(ikx)$. In the case of a crystalline solid, states are specified by n (the band index) and k (the 'crystal momentum').

For a weak potential of period a, $E = \hbar^2 k^2/2m_e$ except when k is close to $n\pi/a$ ($n = \pm 1, \pm 2, \ldots$)—the Brillouin zone boundaries. The reason why deviations always occur in these regions is that a wave function of the form $\psi(x) = \exp(ikx)$ represents a *travelling* wave in the $+x$-direction, of wavelength $2\pi/k$. From the condition for Bragg reflection: $2a \sin \theta = n\lambda$ (where a is the interplane spacing) we might expect the travelling electron wave to be strongly affected in our one-dimensional case (for which $\sin \theta = 1$) when $2a = n\lambda$, i.e. $2a = n \cdot 2\pi/k$, i.e. $k = n\pi/a$. The electron waves cannot propagate at the zone boundaries and a discontinuity occurs in the E–k

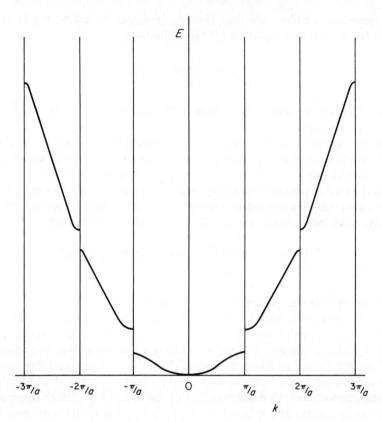

Figure 1. A typical E–k relationship for a weak potential. Discontinuities in E occur at $k = n\pi/a$ ($n = \pm1, \pm2, \ldots$)

relationship (see Figure 1). The region for $0 < |k| < \pi/a$ is called the first Brillouin zone, $\pi/a < |k| < 2\pi/a$ the second Brillouin zone, etc.

While these general arguments give a qualitative picture, a determination of the details of the variation of E with k depends on solving the Schrödinger equation for the particular periodic potential involved.

2.2 Bloch functions

The time-independent Schrödinger equation for an electron in one dimension is

$$\left[-\frac{\hbar^2}{2m_e}\frac{d^2}{dx^2} + V(x) \right]\psi(x) = E\psi(x). \tag{1}$$

If we use atomic units (see Chapter 9) and measure energies in Rydbergs and distances in Bohr radii then this is· equivalent to setting $\hbar = 1$, $m_e = \frac{1}{2}$. Hence for an electron equation (1) takes the form:

$$\left[-\frac{d^2}{dx^2} + V(x) \right] \psi(x) = E\psi(x), \tag{2}$$

where $V(x)$, the potential, is periodic with period a, i.e. $V(x) = V(x + ma)$ and m is an integer.

It might appear that $\psi(x)$ should also be periodic; in fact the reality is more complicated. The probability density $\psi^*(x)\psi(x)$ is indeed periodic with period a, but this can still hold if $\psi(x)$ itself is equal to the product of a function which is periodic ($u(x)$ say) and a complex quantity whose product with its own complex conjugate is unity. The general form of such a complex quantity must be $\exp(if[x, k])$, i.e. $\psi(x) = \exp(if[x, k])u(x)$ and

$$\begin{aligned} \psi^*(x)\psi(x) &= \exp(-if[x, k])u^*(x)\exp(if[x, k])u(x) \\ &= u^*(x)u(x), \end{aligned} \tag{3}$$

which shows that the probability density is of period a. We know that for $V(x) \to 0$, $\psi(x) \to \exp(ikx)$ and so the possibility suggests itself that $\exp(if[x, k]) = \exp(ikx)$ and $\psi(x) = \exp(ikx)u_k(x)$, where the k subscript on $u_k(x)$ implies a dependence of the periodic function on k. Wave functions of this form are known as Bloch functions. (Bloch first established this result in the present connection,[4] on the basis of the periodic potential and 'periodic boundary condition'). One consequence of the form of the Bloch functions is that there are states at $k + 2n\pi/a$ $(n = \pm 1, \pm 2, \ldots)$ with the same energy as a state at k. Consider the state $\psi_k(x) = \exp(ikx)u_k(x)$ and the state at $k' = k + 2n\pi/a$;

$$\begin{aligned} \psi_{k'}(x) &= \exp(ik'x)u_{k'}(x) \\ &= \exp(ikx)\exp(i2n\pi x/a)u_{k'}(x), \end{aligned} \tag{4}$$

where $\exp(i2n\pi x/a)$ is of period a so that $\exp(i2n\pi x/a)u_{k'}(x)$ is of period a. A possible form for this function, which will certainly correspond to a solution of equation (1) is

$$\exp(i2n\pi x/a)u_{k'}(x) = u_k(x), \tag{5}$$

in which case

$$\psi_k(x) = \psi_{k'}(x), \tag{6}$$

i.e. the two states are the same.

There are, in consequence, three entirely equivalent ways in which the E–k relationship can be presented (see Figure 2):

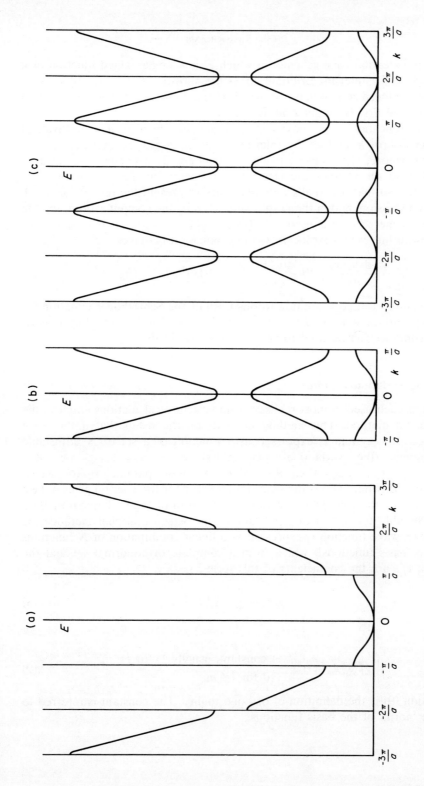

Figure 2. The three equivalent zone schemes: (a) the extended zone scheme; (b) the reduced zone scheme; and (c) the repeated or periodic zone scheme

(a) The extended zone scheme, in which E is a single-valued function of k and all the relevant Brillouin zones are shown.

(b) The reduced zone scheme, in which E is a multivalued function of k and only the first Brillouin zone is shown.

(c) The repeated or periodic zone scheme, in which E is a multivalued function of k, and all the relevant Brillouin zones are shown.

Furthermore, $E(k) = E(-k)$—the electron energy depends on λ but does not depend on whether the electron is travelling in the $+x$- or the $-x$-direction—so that it is really only necessary to present the results for $k > 0$. In this discussion, and in the graphs generated by the computer program, the reduced zone scheme is used with $0 < k < \pi/a$.

Substituting $\psi(x) = \exp(ikx)u_k(x)$ into equation (1) gives

$$-\frac{d^2 u_k(x)}{dx^2} - 2ik\frac{du_k(x)}{dk} + [k^2 + V(x)]u_k(x) = Eu_k(x). \tag{7}$$

It is required to find a solution to this form of the Schrödinger equation for the appropriate $V(x)$. Exact analytic solutions are unobtainable but numerical techniques can be used to achieve accurate results.

2.3 The variational method

The approach used in most realistic band structure calculations employs the variational method. This method depends on the result that if the wave function of a system is expressed as a function of a set of N adjustable parameters (the variational parameters) $c_1, \ldots, c_l, \ldots, c_N$, i.e. $\psi = \psi(x, c_1, \ldots, c_l, \ldots, c_N)$ then the values of these parameters for which $\partial\langle E\rangle/\partial c_l = 0$ (where $\langle E\rangle$ is the expectation value for the energy) gives a 'best estimate' for the true wave function. In the form of the variational method most suitable for numerical applications (the Ritz variational method) the unknown wave function is expressed as a linear combination of N functions χ_l (the 'basis functions') which form a complete orthonormal set and the chosen c_l's are the coefficients of this series, i.e.

$$\psi(x) = \sum_{l=1}^{N} c_l\chi_l(x), \tag{8}$$

where

$$\int \chi_l^*(x)\chi_m(x)\,dx = \begin{cases} \text{constant, usually 1, for } l = m. \\ 0 \text{ for } l \neq m \end{cases} \tag{9}$$

Equation (9) is the definition of orthonormality. The constant is referred to as the 'norm' of the basis functions.

For bound states the limits of this integral (and subsequent integrations in this sub-section) are usually $-\infty$ to $+\infty$. For the non-bound states of interest in band structure work it is usual to carry out the integration over the unit cell ($0 \leqslant x \leqslant a$ here).

For $N \to \infty$ the series approaches an exact solution to the Schrödinger equation. In practice, of course, the series used is finite but is chosen to be sufficiently long that any errors due to truncation are negligibly small. Now $\langle E \rangle$ is given by

$$\langle E \rangle = \frac{\int \psi^*(x) \mathcal{H} \psi(x) \, dx}{\int \psi^*(x) \psi(x) \, dx}. \tag{10}$$

If equation (8) is used then

$$\langle E \rangle \int \sum_{l=1}^{N} c_l^* \chi_l^*(x) \sum_{m=1}^{N} c_m \chi_m(x) \, dx = \int \sum_{l=1}^{N} c_l^* \chi_l^*(x) \mathcal{H} \sum_{m=1}^{N} c_m \chi_m(x) \, dx. \tag{11}$$

The left-hand side can be simplified because of the orthonormality of the functions $\chi_l(x)$ and if we adopt the conventional abbreviation

$$H_{lm} = \int \chi_l^*(x) \mathcal{H} \chi_m(x) \, dx, \tag{12}$$

then

$$\langle E \rangle \sum_{l=1}^{N} c_l^* c_l = \sum_{l=1}^{N} \sum_{m=1}^{N} c_l^* c_m H_{lm}. \tag{13}$$

Taking the derivative with respect to c_l^* gives

$$\frac{\partial \langle E \rangle}{\partial c_l^*} \sum_{l=1}^{N} c_l^* c_l + \langle E \rangle c_l = \sum_{m=1}^{N} c_m H_{lm}. \tag{14}$$

But, according to the variational principle, the value of the c_l's, corresponding to a solution of the Schrödinger equation, occur for

$$\frac{\partial \langle E \rangle}{\partial c_l^*} = 0, \tag{15}$$

which from equation (14) gives

$$\sum_{m=1}^{N} c_m H_{lm} = \langle E \rangle c_l \qquad (l = 1 \ldots N). \tag{16}$$

This constitutes a set of N simultaneous linear equations which have non-trivial solutions if

$$
\begin{vmatrix}
H_{11} - \langle E \rangle & H_{12} & \cdots\cdots\cdots\cdots & H_{1N} \\
H_{21} & H_{22} - \langle E \rangle & & \\
\vdots & & & \vdots \\
& & & \\
& & & \\
& & & \\
H_{N1} & \cdots\cdots\cdots\cdots\cdots\cdots\cdots & & H_{NN} - \langle E \rangle
\end{vmatrix} = 0. \tag{17}
$$

This equation has N roots for $\langle E \rangle$ and can be solved by standard numerical techniques. The N roots constitute a 'best estimate' of the N lowest energy levels of the system, for the chosen basis functions. The corresponding values for the c_i's, and hence $\psi(x)$, can also be found.

A key issue is therefore the choice of a suitable set of basis functions $\chi(x)$ and the central problem of the subsequent analysis is the evaluation of

$$
H_{lm} = \int \chi_l^*(x) \mathcal{H} \chi_m(x) \, dx. \tag{18}
$$

One noteworthy feature of the variational method is that the results for $\langle E \rangle$ are generally more accurate than the results for $\psi(x)$ because a first-order error in $\psi(x)$ only leads to a second-order error in $\langle E \rangle$.

2.4 The basis functions

To apply the Ritz variational method we must express $u_k(x)$ as a linear combination of complete orthonormal functions. The fact that $u_k(x)$ is periodic, of period a, suggests the possibility of expressing $u_k(x)$ as a Fourier series:

$$
u_k(x) = \frac{a_0}{2} + \sum_{l=1}^{N} a_l \cos\left(\frac{2\pi l x}{a}\right) + \sum_{l=1}^{N} b_l \sin\left(\frac{2\pi l x}{a}\right). \tag{19}
$$

(The terms of a Fourier series form a complete orthonormal set.) This approach was adopted in the present work and is useful for a general-purpose treatment. In fact, as we shall see, there are better ways of expanding $u_k(x)$ for the case of realistic potentials.

Rather than employing explicit cosine and sine terms it is more convenient to use $\chi_l(x) = \exp[i(2\pi lx/a)]$ and to express the Fourier series as

$$u_k(x) = \sum_{l=-N}^{+N} c_l \exp\left(\frac{i2\pi lx}{a}\right), \tag{20}$$

where the c_l's are in general complex. These functions are normalized to 'a' rather than 1 over the unit cell, but this factor subsequently cancels.

The potential $V(x)$ is also periodic and can likewise be expanded as a Fourier series:

$$V(x) = \frac{d_0}{2} + \sum_{j=1}^{N} d_j \cos\left(\frac{2\pi jx}{a}\right) + \sum_{j=1}^{N} e_j \sin\left(\frac{2\pi jx}{a}\right). \tag{21}$$

Many crystals possess what is called an 'inversion centre' and if such a point is chosen as the origin of coordinates then $V(\mathbf{r}) = V(-\mathbf{r})$. Symmetry of this kind leads to a considerable reduction in the computing involved and for all the potentials investigated here we can choose an origin such that $V(x) = V(-x)$. Now $\cos(ax) = \cos(-ax)$ but $\sin(ax) = -\sin(-ax)$ so, if we make this choice of origin, all the e_j's are zero. It is again more convenient to write

$$V(x) = \sum_{j=-N}^{+N} f_j \exp\left(\frac{i2\pi jx}{a}\right). \tag{22}$$

Because the sine terms are absent

$$f_j = f_{-j} = \frac{d_j}{2}, \tag{23}$$

and, since $V(x)$ is real, all the f_j's are real.

The Hamiltonian operator takes the form

$$\mathcal{H} = -\frac{d^2}{dx^2} - 2ik\frac{d}{dx} + k^2 + V(x)$$

$$= -\frac{d^2}{dx^2} - 2ik\frac{d}{dx} + k^2 + \sum_{j=-N}^{+N} f_j \exp\left(\frac{i2\pi jx}{a}\right). \tag{24}$$

Hence

$$H_{lm} = \frac{1}{a} \int_0^a \exp\left(\frac{-i2\pi lx}{a}\right)\left[\frac{4\pi^2 m^2}{a^2} + \frac{4\pi mk}{a}\right.$$

$$\left. + k^2 + \sum_{j=-N}^{N} f_j \exp\left(\frac{i2\pi jx}{a}\right)\right]\exp\left(\frac{i2\pi mx}{a}\right) dx. \tag{25}$$

The integration is over the unit cell and the factor of $1/a$ cancels with the norm of the basis functions. Because of the orthogononality of the basis functions over the unit cell, equation (25) becomes

$$H_{lm} = \begin{cases} f_{l-n} & (l \neq m) \\ \left(k + \dfrac{2\pi m}{a} \right)^2 + f_0 & (l = m). \end{cases} \tag{26}$$

Hence, using equation (23) the matrix Hamiltonian is

$$\mathcal{H} = \begin{bmatrix} \left(k - \dfrac{2N\pi}{a} \right)^2 + \dfrac{d_0}{2} & \cdots\cdots\cdots\cdots\cdots\cdots\cdots\cdots\cdots\cdots\cdots\cdots & \dfrac{d_{2N}}{2} \\[2mm] & \left(k - \dfrac{2\pi}{a} \right)^2 + \dfrac{d_0}{2} & \dfrac{d_1}{2} & \dfrac{d_2}{2} \\[2mm] & \dfrac{d_1}{2} & k^2 + \dfrac{d_0}{2} & \dfrac{d_1}{2} \\[2mm] & \dfrac{d_2}{2} & \dfrac{d_1}{2} & \left(k + \dfrac{2\pi}{a} \right)^2 + \dfrac{d_0}{2} \\[2mm] \dfrac{d_{2N}}{2} & \cdots\cdots\cdots\cdots\cdots\cdots\cdots\cdots\cdots\cdots\cdots\cdots & \left(k + \dfrac{2N\pi}{a} \right)^2 + \dfrac{d_0}{2} \end{bmatrix}. \tag{27}$$

The task therefore reduces to Fourier analysing $V(x)$, setting up the Hamiltonian matrix $\mathcal{H}$ (27) and then employing numerical techniques to determine the energies (diagonalization) and, if required, the corresponding wave functions.

2.5 The periodic potential

In a real crystal the periodic potential arises from the Coulomb interaction of the electron with all of the atomic nuclei and all of the other electrons. In the computer program presented here the user can choose between a number of potentials which have been selected principally with a view to their heuristic value rather than to their similarity to real crystal potentials. The potentials are of period a which is set by the user. The origin is an inversion centre, i.e. $V(x) = V(-x)$ so that it is only necessary to specify them for $0 \leqslant x \leqslant a/2$. The potentials are (see Figure 3):

(i) A rectangular potential

$$V(x) = \begin{cases} 0 & 0 \leqslant x < \left(\dfrac{a}{2} - \dfrac{b}{2} \right) \\ V_0 & \left(\dfrac{a}{2} - \dfrac{b}{2} \right) < x \leqslant \dfrac{a}{2}. \end{cases} \tag{28}$$

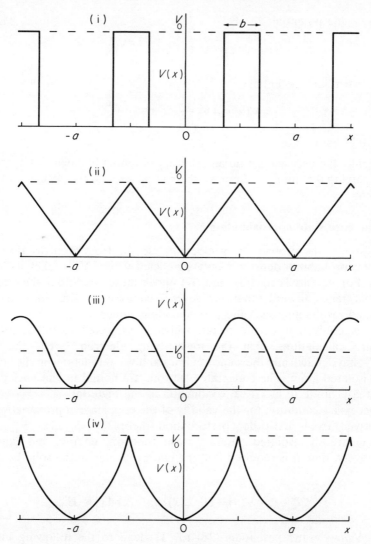

Figure 3. The potential options: (i) the rectangular potential; (ii) the sawtooth potential; (iii) the cosine potential; and (iv) the harmonic potential. (An arbitrary potential can also be set up by linear interpolation)

(ii) A 'sawtooth' potential

$$V(x) = 2V_0 \frac{x}{a}, \qquad 0 < x < \frac{a}{2}. \qquad (29)$$

This form is used so that $V(x = a/2) = V_0$.

(iii) A cosine potential

$$V(x) = V_0\left\{1 - \cos\left(\frac{2\pi x}{a}\right)\right\}. \tag{30}$$

(iv) A 'harmonic' potential

$$V(x) = 4V_0\frac{x^2}{a^2} \qquad 0 < x < \frac{a}{2}, \tag{31}$$

so that again $V(x = a/2) = V_0$.

(v) Finally the user can set up an arbitrary potential by linear interpolation specifying the number of points, their x-values, and the potential at each point.

2.6 The core state approximation

For $V_0 a \to \infty$ the probability of elections moving from atom to atom falls and their probability density is only significant where $V(x)$ takes its lowest values. For potentials (i), (iii), and (iv) under these conditions the effective potential tends towards that for several well-known situations and the computed results approach the corresponding values.

This is equivalent to the result, which is exploited in full-scale band structure calculations, that the lowest-lying electron levels—the 'core states'—have practically the same energies and wave functions in the solid as in the isolated atom. The core state electrons are bound to one nucleus and do not contribute to electrical conduction in the material.

A necessary condition for the validity of the core state approximation can be derived from first-order perturbation theory. If $\phi_{atom}(x)$ E_{atom} and $U_{atom}(x)$ are the isolated 'atomic' wave function, energy, and potential respectively, then it is required that the energy change in the solid be small, i.e.

$$\left|\int_{-\infty}^{+\infty} \phi^*_{atom}(x)[V(x) - U_{atom}(x)]\phi_{atom}(x)\,dx\right| \ll |E_{atom}|. \tag{32}$$

For $V_0 a \to \infty$ the potentials (28) to (31) lead to the following limiting energies and normalized wave functions in atomic units (for $-a/2 < x < a/2$):

(i) For $V_0 > 0$ and $V_0 a \to \infty$ the rectangular potential (28) tends towards an infinite potential well of width $(a - b)$.

$$E_n \to \frac{\pi^2 n^2}{(a-b)^2}, \qquad\qquad n = 1, 2, 3, \ldots, \tag{33}$$

$$\psi_n \to \frac{1}{\sqrt{a-b}}\cos\left(\frac{n\pi x}{a-b}\right), \qquad n = 1, 3, \ldots, \tag{34}$$

$$\psi_n \to \frac{1}{\sqrt{a-b}}\sin\left(\frac{n\pi x}{a-b}\right), \qquad n = 2, 4, \ldots . \tag{35}$$

The situation could be more realistically approximated by a finite potential well of depth V_0, though this problem does not have an analytic solution.[5,6]

(ii) There is no analytic solution for the core state of the sawtooth potential.

(iii) In the case of the cosine potential (30), $V(x)$ is lowest for $x \to 0$ and in this region

$$V_0\left[1-\cos\left(\frac{2\pi x}{a}\right)\right] \to V_0\left[1-\left(1-\frac{4\pi^2 x^2}{2!a^2}+\ldots\right)\right] \tag{36}$$

$$\approx V_0 2\pi^2 \frac{x^2}{a^2}.$$

The situation is therefore essentially equivalent to that for the harmonic potential.

(iv) The harmonic potential (31) will give results for $V_0 a \to \infty$ corresponding to the quantum-mechanical harmonic oscillator, i.e.

$$E_n \to (n+\tfrac{1}{2})\frac{4}{a}\sqrt{V_0}, \tag{37}$$

$$\psi_1(x) \to \left(\frac{4V_0^{\frac{3}{8}}}{a^2\pi^2}\right)^{\frac{1}{8}} \exp\left(-\sqrt{V_0}\cdot\frac{x^2}{a}\right), \tag{38}$$

$$\psi_2(x) \to \left(\frac{2^{\frac{5}{4}}V_0^{\frac{3}{8}}}{a^{\frac{3}{4}}\pi^{\frac{1}{4}}}\right)x \exp\left(-\sqrt{V_0}\frac{x^2}{a}\right), \tag{39}$$

$$\psi_3(x) \to \frac{1}{2^{\frac{1}{4}}}\left(\frac{V_0}{a^2\pi^2}\right)^{\frac{1}{8}}\left(\frac{4}{a}\sqrt{V_0}x^2-1\right)\exp\left(-\sqrt{V_0}\frac{x^2}{a}\right). \tag{40}$$

In all cases, as $V_0 a$ is increased, the lowest energy state will approach the core state limit soonest because, having less energy, it is more closely confined to the regions where $V(x)$ is low.

As an example of the application of the core state approximation the data points in Figure 4 shows the results of a series of runs of the computer program for the energies at $k = 0$ employing the cosine potential for a wide range of values of V_0. The solid lines show the energies of the three lowest energy states of an electron as predicted by the core state approximation (i.e. $E_n = (n+\tfrac{1}{2})(2\pi/a)\sqrt{2V_0}$), for the cosine potential. Note that agreement deteriorates for low V_0, particularly for the third band.

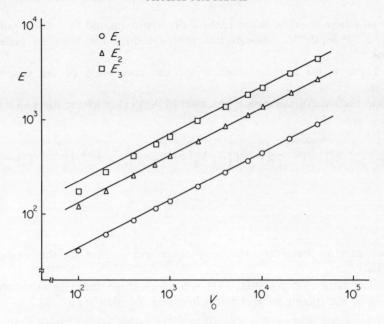

Figure 4. The results of the computer program employing the cosine potential with a period of 1.0 showing the energies at $k = 0$ as a function of V_0: $E_1(\bigcirc)$, $E_2(\triangle)$, and $E_3(\square)$. The straight lines show the predictions of the core state approximation

The computer program also finds the probability density $\psi^*(x)\psi(x)$. In Figure 5 the data points show the computed results for $k = 0$ with the cosine potential; the solid curves are the predictions of the core state approximation (which are simple analytic expressions). Agreement is poorest at larger $|x|$ where the difference between

$$V_0\left\{1 - \cos\left(\frac{2\pi x}{a}\right)\right\} \quad \text{and} \quad V_0\left(\frac{2\pi^2 x^2}{a^2}\right)$$

is greatest.

2.7 The nearly-free electron approximation

The opposite extreme of very low values of $V_0 a$ can be treated using the nearly-free approximation. If $V(x) = 0$ then $\psi(x) = \exp(ikx)$ and $E = k^2$. If $V(x)$ is small we can treat it as a perturbation and from the usual expression

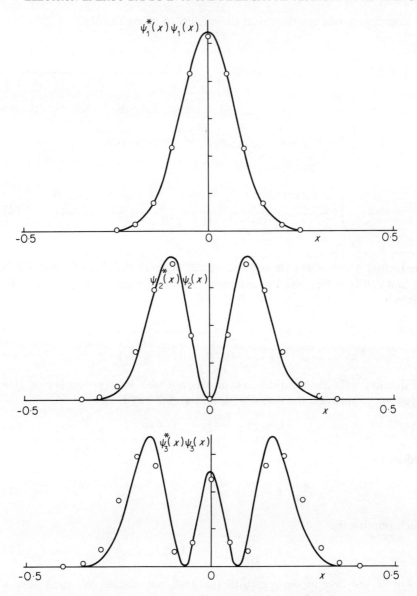

Figure 5. The circles are the results of the computer program for the probability density when the cosine potential is employed with a period of 1.0, $V_0 = 400$ and $k = 0$. The curves show the predictions of the core state approximation

for non-degenerate perturbation theory, up to second order,

$$E = k^2 + \frac{1}{a} \int_0^a \psi_k^*(x) V(x) \psi(x)\, dx + \frac{1}{a^2} \int_{\substack{-\infty \\ k' \neq k}}^{+\infty} \frac{\left| \int_0^a \psi_{k'}^*(x) V(x) \psi_k(x)\, dx \right|^2 dk'}{(k^2 - k'^2)}$$

$$= k^2 + \frac{1}{a} \int_0^a \exp(-ikx) \sum_{j=-N}^{+N} f_j \exp\!\left(\frac{i2\pi jx}{a}\right) \exp(ikx)\, dx$$

$$+ \frac{1}{a^2} \int_{\substack{-\infty \\ k' \neq k}}^{+\infty} \frac{\left| \int_0^a \exp(-ik'x) \sum_{j=-N}^{+N} f_j \exp\!\left(\frac{i2\pi jx}{a}\right) \exp(ikx)\, dx \right|^2 dk'}{(k^2 - k'^2)}. \tag{41}$$

The factors $1/a$ and $1/a^2$ arise because the norm of the wave functions over the unit cell is a. Because the exponential functions are orthogonal over the unit cell,

$$E = k^2 + f_0 + \sum_{\substack{j=-N \\ j \neq 0}}^{+N} \frac{f_j^2}{\left\{ k^2 - \left(k - \frac{2\pi j}{a} \right)^2 \right\}}. \tag{42}$$

Equation (42) is accurate except when there is degeneracy, or near degeneracy, between the states at k and at $k - 2\pi j/a$, i.e.

$$|k| \approx \left| k - \frac{2\pi j}{a} \right|, \qquad (j \neq 0), \tag{43}$$

leading to

$$k \approx -k + \frac{2\pi j}{a}, \tag{44}$$

which implies that

$$k \approx \frac{\pi j}{a} \qquad (j = \pm 1, \pm 2, \ldots), \tag{45}$$

as will be the case in the region of the zone boundaries. We then need to consider the explicit form of the wave functions.

As an example of the approach consider the lowest energy band. The unperturbed wave function $\psi(x) = \exp(ikx)$. The perturbation $V(x)$ will 'mix in' states with wave functions

$$\exp\!\left\{ i\!\left(k - \frac{2\pi j}{a} \right) x \right\} \qquad j = \pm 1, \pm 2 \ldots$$

but, close to $k = \pi/a$ (the zone boundary) the main contribution will be due to the state $\exp\{i(k - 2\pi/a)x\}$ that is degenerate with the unperturbed state at the zone boundary, i.e.

$$\psi(x) \approx \alpha \exp(ikx) + \beta \exp\left\{i\left(k - \frac{2\pi}{a}\right)x\right\}, \tag{46}$$

where α and β are to be determined. The Schrödinger equation takes the form:

$$\left[-\frac{d^2}{dx^2} + \sum_{j=-N}^{+N} f_j \exp\left(\frac{i2\pi j}{a}\right)\right]\left(\alpha \exp(ikx) + \beta \exp\left\{i\left(k - \frac{2\pi}{a}\right)x\right\}\right)$$

$$= E\left(\alpha \exp(ikx) + \beta \exp\left\{i\left(k - \frac{2\pi}{a}\right)x\right\}\right), \tag{47}$$

that becomes

$$(k^2 - E)\alpha \exp(ikx) + \left[\left(k - \frac{2\pi}{a}\right)^2 - E\right]\beta \exp\left\{i\left(k - \frac{2\pi}{a}\right)x\right\}$$

$$+ \sum_{j=-N}^{+N} f_j \exp\left(\frac{i2\pi j}{a}\right)\left[\alpha \exp(ikx) + \beta \exp\left\{i\left(k - \frac{2\pi}{a}\right)x\right\}\right] = 0. \tag{48}$$

Since this equation is true for all x the coefficients of all the exponential terms must vanish. The coefficient of $\exp(ikx)$ is

$$(k^2 - E)\alpha + f_0\alpha + f_1\beta = 0, \tag{49}$$

and of $\exp\left\{i\left(k - \frac{2\pi}{a}\right)x\right\}$ is

$$\left[\left(k - \frac{2\pi}{a}\right)^2 - E\right]\beta + f_0\beta + f_{-1}\alpha = 0. \tag{50}$$

The other exponential terms should strictly be incorporated into similar equations involving the other (negligible) admixed wave functions. Eliminating α/β from the two equations above and substituting for f_0 and $f_1(= f_{-1})$ gives

$$\left[k^2 + \frac{d_0}{2} - E\right]\left[\left(k - \frac{2\pi}{a}\right)^2 + \frac{d_0}{2} - E\right] - \frac{d_1^2}{4} = 0, \tag{51}$$

which has the roots

$$E = k^2 + \frac{d_0}{2} + \frac{2\pi}{a}\left[\frac{\pi}{a} - k \pm \sqrt{\left(\frac{\pi}{a} - k\right)^2 + \left(\frac{ad_1}{4\pi}\right)^2}\right]. \tag{52}$$

The lower root applies to the lowest energy band. The higher root applies to the second energy band close to the first zone boundary.

The energy gap at the first zone boundary ($k = \pi/a$) is therefore $|d_1|$, and quite generally the energy gap at the jth zone boundary is $|d_j|$.

This result can be readily obtained by diagonalizing the degenerate two-dimensional submatrix of the Hamiltonian matrix (27) appropriate at the jth zone boundary, i.e.

$$
\begin{bmatrix}
\dfrac{\pi^2 j^2}{a^2} + \dfrac{d_0}{2} & \dfrac{d_j}{2} \\[2ex]
\dfrac{d_j}{2} & \dfrac{\pi^2 j^2}{a^2} + \dfrac{d_0}{2}
\end{bmatrix}
$$

Hence

$$
E(k = \pi j/a) = \frac{\pi^2 j^2}{a^2} + \frac{d_0}{2} \pm \frac{d_j}{2}. \tag{53}
$$

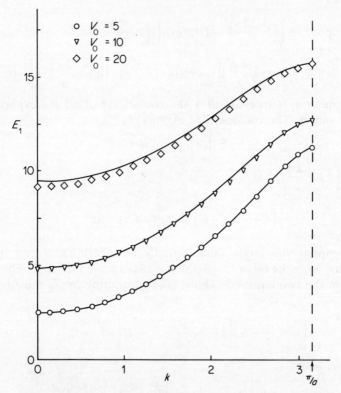

Figure 6. The results for the E–k relationship of the lowest band found by the computer program with a sawtooth potential of period 1.0 for $V_0 = 5(\bigcirc)$, $V_0 = 10(\triangledown)$ and $V_0 = 20(\diamondsuit)$. The solid curves show the predictions of the nearly-free electron approximation

Figure 6 shows a comparison between the results of the computer program (the data points) and the nearly-free electron approximation (the solid curves) for the $E-k$ relationship of the lowest band for three values of V_0 with the sawtooth potential. In this case agreement deteriorates for *high* V_0. Figure 7 shows the results of a series of program runs for the energy gap at the first zone boundary (the circles) as a function of V_0, for the sawtooth potential. The solid line shows the prediction of the nearly-free electron approximation, i.e. $4V_0/\pi^2$. Agreement decreases at large V_0.

It is of some interest to determine the coefficients α and β, and hence the wave function, especially at the zone boundary. Now using equations (23) and (49)

$$\alpha\left(k^2 + \frac{d_0}{2} - E\right) = -\beta\frac{d_1}{2},$$ (54)

and, at the zone boundary.

$$E = k^2 + \frac{d_0}{2} \pm \frac{d_1}{2}.$$ (55)

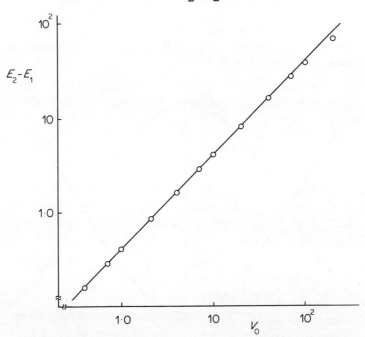

Figure 7. The energy gap at the first zone boundary: $E_2(k = \pi/a) - E_1(k = \pi/a)$ as a function of V_0. The circles show the results of the computer program for a sawtooth potential of period 1.0. The solid line shows the prediction of the nearly-free electron approximation

If $d_1 < 0$, which is the case here for $V_0 > 0$, then for the lowest energy band the upper sign in equation (55) is taken and

$$\alpha\left(-\frac{d_1}{2}\right) = -\beta\left(\frac{d_1}{2}\right). \tag{56}$$

The normalization requires $\alpha^2 + \beta^2 = 1$ so that for the lowest band, at the zone boundary,

$$\psi(x) = \frac{1}{\sqrt{2}}\left[\exp\left(\frac{i\pi x}{a}\right) + \exp\left(-\frac{i\pi x}{a}\right)\right]$$

$$= \sqrt{2}\cos\left(\frac{\pi x}{a}\right). \tag{57}$$

For the second band at $k = \pi/a$, a similar analysis gives

$$\psi(x) = i\sqrt{2}\sin\left(\frac{\pi x}{a}\right). \tag{58}$$

Thus, for both bands, the wave functions at the zone boundary correspond to standing waves (due to Bragg reflection) rather than travelling waves. Furthermore, the probability density is high in the region of low potential for the lower energy band and high in the region of high potential for the higher band.

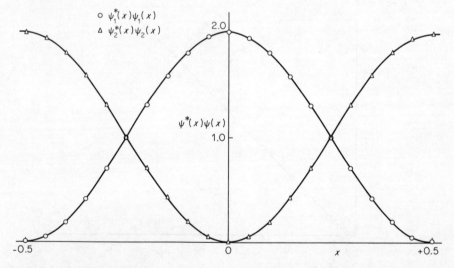

Figure 8. The results of the computer program for the probability density at $k = \pi/a$ of the first (O) and the second band ($\triangle$) for a sawtooth potential of period 1.0 with $V_0 = 1.0$. The solid curves show the predictions of the nearly-free electron approximation. Note that for the lower energy band the probability density is high where $V(x)$ is low; for the higher energy band the probability density is high where $V(x)$ is high

In Figure 8 the computed results for the probability density (the data points) are compared with the above results for the nearly-free electron approximation. The potential used was the sawtooth potential.

In view of the rather large magnitude of the Fourier components of the Coulomb potential it might be thought that the nearly-free electron approximation would be of little value in practice, and indeed, for many years, it was not clear why it gives reasonable results in many cases. The reason is that, because the 'free' electron wave functions are orthogonal to the core states, the Schrödinger equation can be transformed so that the potential can be replaced by a 'pseudopotential'[7] which is weaker than the true Coulomb potential and which has Fourier components of smaller magnitude which are rapidly convergent.

2.8 The tight-binding approximation

If $V_0 a$ is large, but not sufficiently large for the core state approximation to apply, we might expect that a good approximation to the true wave function would be a linear combination of atomic wave functions $\phi_{\text{atom}}(x)$ on each site. Bloch[4] suggested a linear combination of the form:

$$\psi_k(x) = A \sum_{n=-\infty}^{+\infty} \exp(ikna)\phi_{\text{atom}}(x-na), \tag{59}$$

where $\phi_{\text{atom}}(x-na)$ is an atomic wave function centred on the site at na, and A is a normalizing factor.

These functions satisfy the Bloch condition $\psi_k(x) = \exp(ikx)u_k(x)$, where $u_k(x)$ is of period a. The suggested linear combination implies

$$u_k(x) = \exp(-ikx)A \sum_{n=-\infty}^{+\infty} \exp(ikna)\phi_{\text{atom}}(x-na). \tag{60}$$

Hence

$$u_k(x+ma) = \exp\{-ik(x+ma)\}A \sum_{n=-\infty}^{+\infty} \exp(ikna)\phi_{\text{atom}}(x+ma-na)$$

$$= \exp(-ikx)A \sum_{n=-\infty}^{+\infty} \exp(ik[n-m]a)\phi_{\text{atom}}(x-[n-m]a), \tag{61}$$

and, since the sum over n runs from $-\infty$ to $+\infty$,

$$u_k(x+ma) = \exp(-ikx)A \sum_{n=-\infty}^{+\infty} \exp(ikna)\phi_{\text{atom}}(x-na)$$

$$= u_k(x). \tag{62}$$

Now equation (10) is

$$\langle E \rangle = \frac{\int_0^a \psi^*(x)\mathcal{H}\psi(x)\,dx}{\int_0^a \psi^*(x)\psi(x)\,dx} = \frac{N}{D}. \tag{63}$$

The denominator, using equation (60), is

$$D = \int_0^a \psi^*(x)\psi(x)\,dx$$

$$= A^*A \int_0^a \sum_{n=-\infty}^{+\infty} \exp(-ika)\phi^*_{\mathrm{atom}}(x-na) \sum_{m=-\infty}^{+\infty} \exp(ikma)\phi_{\mathrm{atom}}(x-ma)\,dx. \tag{64}$$

If we write $l = n - m$ equation (64) becomes

$$D = A^*A \sum_{n=-\infty}^{+\infty} \sum_{l=-\infty}^{+\infty} \exp(-kla) \int_0^a \phi^*_{\mathrm{atom}}(x-na)\phi_{\mathrm{atom}}(x+la-na)\,dx. \tag{65}$$

Summing over n and then integrating over the interval $0 \leqslant x \leqslant a$, i.e. over the unit cell, is equivalent to integrating over all space giving

$$D = A^*A \sum_{l=-\infty}^{+\infty} \exp(-ikla) \int_{-\infty}^{+\infty} \phi^*_{\mathrm{atom}}(x)\phi_{\mathrm{atom}}(x+la)\,dx. \tag{66}$$

If we write

$$S_l - \int_{-\infty}^{+\infty} \phi^*_{\mathrm{atom}}(x)\phi_{\mathrm{atom}}(x+la)\,dx, \tag{67}$$

equation (66) simplifies to

$$D = A^*A \sum_{l=-\infty}^{+\infty} \exp(-ikla)S_l. \tag{68}$$

(Note that S_0 is simply the norm of $\phi_{\mathrm{atom}}(x)$.)

Major simplification occurs because, when the tight-binding approximation is appropriate, the atomic wave functions $\phi_{\mathrm{atom}}(x)$ are highly localized, and S_l is significant only for $l = 0$ (the dominant term) and for $l = \pm 1$.

The numerator in the expression for $\langle E \rangle$ is

$$N = \int_0^a \psi^*(x)\mathcal{H}\psi(x)\,dx$$

$$= AA^* \int_0^a \sum_{n=-\infty}^{+\infty} \exp(-ikna)\phi_{atom}^*(x-na)\mathcal{H} \sum_{m=-\infty}^{+\infty}$$

$$\times \exp(ikma)\phi_{atom}(x-ma)\,dx, \quad (69)$$

and, by a similar argument, if we define

$$H_l = \int_{-\infty}^{+\infty} \phi_{atom}^*(x)\mathcal{H}\phi_{atom}(x+la)\,dx, \quad (70)$$

equation (69) becomes

$$N = A^*A \sum_{l=-\infty}^{+\infty} \exp(-ikla)H_l. \quad (71)$$

Now

$$\mathcal{H}\phi_{atom}(x+la) = \left[-\frac{d^2}{dx^2} + V(x)\right]\phi_{atom}(x+la)$$

$$= \left[-\frac{d^2}{dx^2} + U_{atom}(x+la)\right]\phi_{atom}(x+la)$$

$$+ [V(x) - U_{atom}(x+la)]\phi_{atom}(x+la), \quad (72)$$

(here $U_{atom}(x+la)$ is the atomic potential centred on the atom at $-la$). Therefore,

$$\mathcal{H}\phi_{atom}(x+la) = E_{atom}\phi_{atom}(x+la) + [V(x) - U_{atom}(x+la)]\phi_{atom}(x+la), \quad (73)$$

so that

$$H_l = \int_{-\infty}^{+\infty} \phi_{atom}^*(x)\mathcal{H}\phi_{atom}(x+la)\,dx$$

$$= E_{atom}S_l + g_l, \quad (74)$$

where

$$g_l = \int_{-\infty}^{+\infty} \phi_{atom}^*(x)[V(x) - U_{atom}(x+la)]\phi_{atom}(x+la)\,dx. \quad (75)$$

The g_l's are referred to as overlap integrals. Hence

$$\langle E \rangle = \frac{\sum\limits_{l=-\infty}^{+\infty} \exp(-ikla) \cdot (E_{\text{atom}} S_l + g_l)}{\sum\limits_{l=-\infty}^{+\infty} \exp(-ikla) S_l}$$

$$= E_{\text{atom}} + \frac{\sum\limits_{l=-\infty}^{+\infty} \exp(-ikla) g_l}{\sum\limits_{l=-\infty}^{+\infty} \exp(-ikla) S_l}, \tag{76}$$

and the only significant contributions will come from terms for small $|l|$.

Figure 9 shows some of the functions involved in the integrals which give S_l and g_l for the case of the lowest energy band with the harmonic potential. For $V_0 a \to \infty$ the only significant terms are for $l = 0$ and

$$\langle E \rangle \approx E_{\text{atom}} + \frac{g_0}{S_0}. \tag{77}$$

The first term is the core state approximation. The second term is the correction to it predicted by first-order perturbation theory.

For somewhat lower values of $V_0 a$ the $|l| = 1$ terms will also be significant.

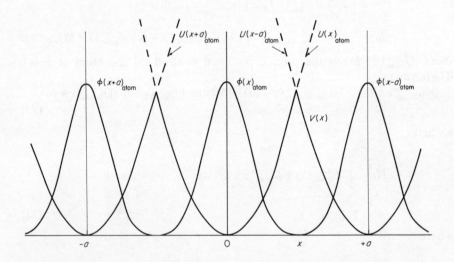

Figure 9. Some of the functions involved in calculating S_l and g_l for the lowest energy band in the tight-binding approximation with the harmonic potential

Usually the term in $g_{\pm 1}$ is significant before that in $S_{\pm 1}$ and, since $g_1 = g_{-1}$,

$$\langle E \rangle \approx E_{\text{atom}} + \frac{g_0}{S_0} + \frac{g_1}{S_0}[\exp(ika) + \exp(-ika)]$$

$$= E_{\text{atom}} + \frac{g_0}{S_0} + \frac{2g_1}{S_0}\cos(ka), \tag{78}$$

and, if $\phi_{\text{atom}}(x)$ is normalized to unity,

$$\langle E \rangle = E_{\text{atom}} + g_0 + 2g_1 \cos(ka). \tag{79}$$

Thus, in the tight-binding approximation,

$$E(k=0) - E(k=\pi/a) = 4g_1 \quad \text{and} \quad E(k=\pi/2a) = E_{\text{atom}} + g_0.$$

The tight-binding approximation can be applied to the following potentials employed in the computer program:

(i) The cosine potential where, for the lowest energy band,

$$g_0 = V_0\left(\frac{2V_0}{a^2}\right)^{\frac{1}{4}} \int_{-\infty}^{+\infty} \exp\left(-\frac{x^2\pi}{a}\sqrt{2V_0}\right)\left[1 - \cos\left(\frac{2\pi x}{a}\right) - \frac{2\pi^2 x^2}{a^2}\right] dx$$

$$= V_0\left[1 - \frac{\pi}{a\sqrt{2V_0}} - \exp\left(-\frac{\pi}{a\sqrt{2V_0}}\right)\right], \tag{80}$$

and

$$g_1 = V_0\left(\frac{2V_0}{a^2}\right)^{\frac{1}{4}} \int_{-\infty}^{+\infty} \exp\left(-\frac{x^2\pi}{a}\sqrt{\frac{V_0}{2}}\right)\left[1 - \cos\left(\frac{2\pi x}{a}\right) - \frac{2\pi^2}{a^2}(x+a)^2\right]$$

$$\times \exp\left(-\frac{[x+a]^2\pi}{a}\sqrt{\frac{V_0}{2}}\right) dx$$

$$= V_0 \exp\left(\frac{-a\pi}{2}\sqrt{\frac{V_0}{2}}\right)\left[1 - \frac{\pi^2}{2} - \frac{\pi}{a\sqrt{2V_0}} + \exp\left(-\frac{\pi}{a\sqrt{2V_0}}\right)\right]. \tag{81}$$

(ii) The harmonic potential. For the lowest band

$$g_0 \approx \frac{8}{a^2}\left(\frac{4V_0}{a^2\pi^2}\right)^{\frac{1}{4}} \int_{a/2}^{+\infty} \exp\left(-2\sqrt{V_0}\frac{x^2}{a}\right)[V_0(x-a)^2 - V_0 x^2] dx$$

$$= 4V_0\left[1 - \Phi(V_0^{\frac{1}{4}}\sqrt{a})\right] - \frac{4\sqrt{2}V_0^{\frac{3}{4}}}{\sqrt{\pi}a}\exp\left(-\sqrt{V_0}\frac{a}{2}\right), \tag{82}$$

$$g_1 \approx \frac{4}{a^2}\left(\frac{4V_0}{a^2\pi^2}\right)^{\frac{1}{4}} \int_{-a/2}^{+a/2} \exp\left(-\frac{\sqrt{V_0}x^2}{a}\right)$$

$$\times [V_0 x^2 - V_0(x+a)^2]\exp\left(-\sqrt{V_0}\frac{(x+a)^2}{a}\right) dx$$

$$= 2\sqrt{2}\cdot\frac{V_0^{\frac{3}{4}}}{\sqrt{\pi a}}\exp\left(-\sqrt{V_0}\frac{a}{2}\right)[\exp(-2\sqrt{V_0}a)-1]. \tag{83}$$

For the harmonic potential the probability integral

$$\Phi(z) = \sqrt{\frac{2}{\pi}}\int_0^z \exp(-t^2/2)\,dt \tag{84}$$

is required. This integral cannot be expressed in closed form. It can be approximated by a series expansion, but for the large values of z typically involved, convergence is slow and it is more convenient to use tabulated values for $\Phi(z)$.

The tight-binding approximation cannot be applied to the rectangular potential if the 'atomic' wave functions are taken to be those corresponding to an infinite potential well because in that case $g_0 = g_1 = \ldots = 0$. It can be used if the atomic states are taken as those appropriate for a finite potential well, and an approach along these lines has been presented by Wetsel[6] though the overlap integrals for the finite potential well cannot be expressed as analytic functions but involve the solution of a transcendental equation.

In Figure 10 the computed results for the variation of E with k of the lowest band (the data points) are compared with the predictions of the tightbinding approximation (the solid curves) for the harmonic potential. Agreement deteriorates as V_0 increases. In Figure 11 the circles show the computed results for $E(k=\pi/a)$ $E(k=0)$ as a function of V_0 with the harmonic potential. The solid curve shows $4g_1$, i.e. the energy difference predicted by the tight-binding approximation.

2.9 The effective mass

There are situations, particularly those involving the dynamic behaviour of the electrons, when the $E-k$ relationship is not the most useful form in which to present the results of a band structure calculation. The wave function $\psi(x) = \exp(ikx)u_k(x)$ extends over all space; if we want to represent the motion of a 'localized' electron the uncertainty principle indicates that we must build up a wave-packet with a spread of k-values. The appropriate velocity is then the group velocity $(^{\bullet}V_G)$ which is equal to the derivative of

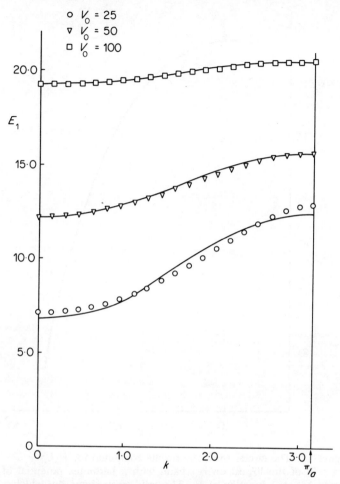

Figure 10. The results of the computer program for the lowest energy band with the harmonic potential of period 1.0 for $V_0 = 25(\bigcirc)$, $V_0 = 50(\bigtriangledown)$, and $V_0 = 100(\square)$. The solid curves show the predictions of the tight-binding approximation

the angular frequency (ω) with respect to k

$$V_G = \frac{\partial \omega}{\partial k} = \frac{\partial E}{\partial k} \tag{85}$$

($E = \omega$, since $\hbar = 1$ in the atomic system of units).

If a force F acts on this electron wave-packet in the $+x$-direction then

$$FV_G = \frac{\partial E}{\partial t}, \tag{86}$$

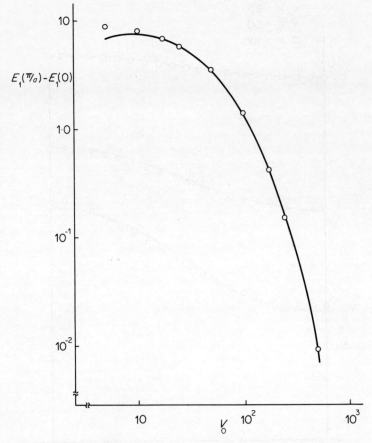

Figure 11. The circles show the results computed for $E(k = \pi/a) - E(k = 0)$ of the lowest energy band with a harmonic potential of period 1.0, as a function of V_0. The solid curve shows $4g_1$ which is the energy difference predicted by the tight-binding approximation

and

$$\frac{\partial V_G}{\partial t} = \frac{\partial}{\partial t}\left(\frac{\partial E}{\partial k}\right) = \frac{\partial}{\partial k}\left(\frac{\partial E}{\partial t}\right)$$

$$= \frac{\partial}{\partial k}(FV_G) = \frac{\partial}{\partial k}\left(F\frac{\partial E}{\partial k}\right)$$

$$= F\frac{\partial^2 E}{\partial k^2}. \tag{87}$$

Comparing this result with Newton's second law of motion for a particle of mass M, i.e. $\partial V/\partial t = (1/M)F$ we see that the motion of the electron wave-packet can be described by the semi-classical concept of an effective mass $(m^*) = 1/(\partial^2 E/\partial k^2)$.

We should perhaps observe that there are situations[8,9] where a different definition of effective mass, $1/m^* = (1/k)\partial E/\partial k$, appears to be appropriate.

For the lowest band $\partial^2 E/\partial k^2 > 0$ near $k = 0$ but $\partial^2 E/\partial k^2 < 0$ near $k = \pi/a$. Rather than refer to a negative effective mass it is conventional to take $m^* = 1/|\partial^2 E/\partial k^2|$ but to refer to electrons (with electronic charge $-e$) when $\partial^2 E/\partial k^2 > 0$ and holes (with charge $+e$)—regarded as the *absence* of an electron from a band—for $\partial^2 E/\partial k^2 < 0$.

It should be noted that, since in the atomic system of units the mass of an electron $m_e = \frac{1}{2}$,

$$\frac{m^*}{m_e} = 2 \Big/ \left|\frac{\partial^2 E}{\partial k^2}\right| \quad \text{(in atomic units)}. \tag{88}$$

It is easy to see that equation (88) is plausible in the two extremes of completely free electrons and the core state approximation. For completely free electrons $E = k^2$, $\partial^2 E/\partial k^2 = 2$ and $m^* = m_e$. In the case of the core state approximation (where the electrons are not free to travel through the crystal) E is independent of k, $\partial^2 E/\partial k^2 \to 0$, and $m^* \to \infty$.

If the nearly-free electron approximation or the tight-binding approximation apply, then the appropriate analytic expressions for $E(k)$ can be transformed to analytic expressions for m^*/m_e as a function of k.

The seemingly perverse behaviour of holes can be understood if we recall that at the zone boundaries Bragg reflection leads to wave functions which are standing waves. Consequently, a wave-packet at a boundary has $V_G = 0$ and cannot propagate. For the lowest energy band then, at $k = 0$ the wave-packet has zero velocity; as k and E increase its velocity increases initially, however, when $k = \pi/a$ the velocity has fallen again to zero. Consequently, there are regions of the energy bands approaching the zone boundaries where an *increase* in energy leads to a *decrease* in velocity—the acceleration is in the opposite direction to the force. The concept of holes is used to describe this unexpected behaviour.

Having found E at $k = 0$, $\pi/20a$, $\pi/10a$, ..., π/a the program estimates $\partial^2 E/\partial k^2$ at these k-values from the results for $E_{k-\Delta k}$, E_k, and $E_{k+\Delta k}$. By Taylor's series:

$$E_{k+\Delta k} = E_k + \frac{\Delta k}{1!}\frac{\partial E}{\partial k} - \frac{\Delta k^2}{2!}\frac{\partial^2 E}{\partial k^2} + \cdots . \tag{89}$$

Hence

$$\left(\frac{\partial^2 E}{\partial k^2}\right)_k \approx \frac{1}{\Delta k^2}[E_{k+\Delta k} + E_{k-\Delta k} - 2E_k]. \tag{90}$$

The program displays the results for m^*/m_e as a function of k, printing a letter E (electron) when $\partial^2 E/\partial k^2 > 0$ and a letter H (hole) when $\partial^2 E/\partial k^2 < 0$.

2.10 The probability density

The variation of E with k (or related variables) is usually the most important result of a band structure calculation. There are, however, situations when the actual wave functions or related quantities are of interest. In particular the total electron probability density, which can be investigated by X-ray and neutron diffraction techniques, according to equation (3) involves the product $u_k^*(x)u_k(x)$. Now

$$u_k(x) = \sum_{l=-N}^{+N} c_l \exp\left(\frac{i2\pi lx}{a}\right), \tag{91}$$

and for systems with an inversion centre, the Hamiltonian matrix is real. Consequently the c_i's are real, and

$$u_k^*(x) = \sum_{m=-N}^{+N} c_m \exp\left(-\frac{i2\pi mx}{a}\right). \tag{92}$$

Hence

$$
\begin{aligned}
u_k^*(x)u_k(x) &= \sum_{m=-N}^{+N} c_m \exp\left(-\frac{i2\pi mx}{a}\right) \sum_{l=-N}^{+N} c_l \exp\left(\frac{i2\pi lx}{a}\right) \\
&= \sum_{m=-N}^{+N} \sum_{l=-N}^{+N} c_m c_l \exp\left(i2\pi[l-m]\frac{x}{a}\right) \\
&= \sum_{m=-N}^{+N} \sum_{l=-N}^{+N} c_m c_l \cos\left(2\pi[l-m]\frac{x}{a}\right) \\
&= 1 + 2 \sum_{m=-N}^{+N} \sum_{l=m+1}^{+N} c_m c_l \cos\left(2\pi[l-m]\frac{x}{a}\right), \tag{93}
\end{aligned}
$$

since $\sum_{m=-N}^{+N} c_m^2 = 1$ and $u_k^*(x)u_k(x)$ is real.

The computer program finds the probability densities in the range $-a/2 < x < a/2$ for a value of k specified by the user. The results can be compared with the predictions of the core state approximation or the nearly-free electron approximation when they are appropriate.

As is generally the case when the variational method is employed, the results for the probability density can be expected to be less accurate than the results for the energy.

3. THE COMPUTER PROGRAM

The program allows the user to choose the potential from the set of even periodic functions described in section 2.5. The first section prints questions about the potential. The program as written is for use on an interactive system. This section would need to be modified for use in batch mode. The user's responses define the potential function by setting the integer variable NPOT and the value of constants, such as the period and the maximum value of the potential.

The subroutine FRANCS(NTERMS, NDATA, DC, A) is then called to find the coefficients of the cosine terms in the Fourier series. The first parameter NTERMS ($= 33$) is the number of coefficients required. NDATA ($= $NPTS) is used only when the interpolated potential is employed. For that particular case FRANCS( ) employs trapezoidal integration and NDATA specifies the number of subdivisions required. The fourth parameter DC is the constant term in the Fourier series for $V(x)$, i.e. $d_0/2$. The fifth parameter is the array A( ) which is set by FRANCS( ) to the Fourier coefficients; $A(1) = d_1$, $A(2) = d_2, \ldots, (A33) = d_{33}$. In the case of all except the interpolated potential the Fourier coefficients are determined analytically.

Next $K(= k)$ is set. It is increased in 20 steps from 0 to π/a. Because the true matrix for the Hamiltonian is of infinite dimension in practice it has to be truncated, which introduces some error into the results. In an attempt to maintain this error less than ~ 1 per cent the dimension of the Hamiltonian (NAR) is set initially to either 10 or 11. The energies are then determined by the subroutine ENERGY( ) and NAR increased by 2. ENERGY( ) is called again and if the fractional changes in the results for the energy are less than 3×10^{-3} then they are printed out. If one or more of the results has changed by more than 3×10^{-3}, NAR is increased by a further 8 and ENERGY( ) called again.

This process is continued till NAR is 32 or 33. At this point the computing time taken becomes significant, so rather than increasing NAR further, the results are printed out with an accuracy warning. If higher accuracy were to be required the program could be modified by increasing the upper limit on NAR.

The subroutine ENERGY(K, E, EV, ND, N, DC, A, H, W, BOOL) is used to find the energy eigenvalues and also (if BOOL is set.FALSE.) the eigenvectors.

The parameters set on entry are: $K(= k)$, ND—the first dimension of H( ) (the Hamiltonian) as specified in ENERGY( ), N is the order of H( ), DC and A( ) are the Fourier coefficients described above and BOOL is a logical variable. If BOOL is set .TRUE. the NAG[10] library routine F02AAF()—which calculates eigenvalues—is called. If BOOL is set .FALSE. the NAG routine F02ABF( )—which calculates eigenvalues and eigenvectors—is called.

ENERGY( ) sets up the Hamiltonian matrix H( ). The input variables common to F02AAF( ) and F02ABF( ) are H( ), ND, N (as described above), and IFAIL which specifies what will happen if a failure occurs during a call to these routines. Both F02AAF and F02ABF require H( ) to be real and symmetric. The array W( ) is used by the NAG routines as working space. On exit from F02AAF or F02ABF E( ) contains the eigenvalues *in ascending order*. On exit from F02ABF the array EV( ) contains the *normalized* eigenvectors in columns corresponding to the eigenvalues. By calling ENERGY( ) for varying values of k the program calculates and prints out results for the E–k diagram, showing the three lowest energy bands.

The user has the option of plotting these results on the line printer. The plot is done by the subroutine GRAPH(Y, NP, NG, X). Y( ) contains the energy values and X( ) the corresponding k-values. If a more sophisticated graph-plotting routine were to be available to the reader GRAPH( ) could be replaced.

The user may also calculate the approximate effective mass as a function of k and the probability density function as a function of x (x being increased in 20 steps from $-a/2$ to $+a/2$), for a chosen value of k. For the probability density function calculation ENERGY( ) is called with BOOL = .FALSE. so that F02ABF( ) is now called. The potential is also found as a function of x by the function V(Y) and printed out with the corresponding probability density.

The calculations of the probability density and the effective mass will be less accurate than the eigenvalue calculations, especially the effective mass results which are only intended to give a qualitative picture of the variation of m^*/m_e with k. In both cases there is the option of plotting the results graphically.

Finally the user has the option of either stopping or of changing the potential and initiating a further set of calculations.

Because of the necessary truncation of the Hamiltonian matrix there is always some error. It is found that significant errors (accompanied by a warning message) generally occur in the extreme core-state limit. This is only to be expected because the very large potential values involved generate large—and non-negligible—Fourier terms, even for the higher terms in the series. For example, the jth coefficient for the harmonic potential is

$$d_j = \frac{2V_0a\cos(j\pi)}{(j\pi)^2}$$

The magnitude of these Fourier terms decreases as j increases but increases with increasing V_0a.

3.1 A typical session

If the rectangular potential is used with:

$$V_0 = 5.0 \qquad\qquad \text{(in Rydbergs)}$$

Period $(a) = 1.5$ (in Bohr radii)

Width of rectangle $(b) = 0.5$ (in Bohr radii)

and if computation of the effective mass and the probability density are requested then the results of the computer program are:

```
PROGRAM TO CALCULATE FIRST THREE ENERGY LEVELS OF
AN ELECTRON SUBJECT TO A GIVEN PERIODIC POTENTIAL.
ALL INPUT AND OUTPUT IS IN ATOMIC UNITS
I.E. THE UNIT OF DISTANCE IS ONE BOHR RADIUS
THE UNIT OF ENERGY IS ONE RYDBERG

WHAT SORT OF POTENTIAL DO YOU WANT
TYPE 1 FOR RECTANGULAR POTENTIAL
TYPE 2 FOR SAWTOOTH POTENTIAL
TYPE 3 FOR COSINE POTENTIAL
TYPE 4 FOR HARMONIC POTENTIAL
TYPE 5 FOR INTERPOLATED POTENTIAL
1
RECTANGULAR POTENTIAL
INPUT HEIGHT OF RECTANGLE AS REAL NUMBER
5.0
INPUT WIDTH OF RECTANGLE AS REAL NUMBER
0.5
INPUT PERIOD AS REAL NUMBER
1.5


        PERIOD        1.5

        K          E1        E2        E3
       0.0000    1.4499   18.4804   20.0641
       0.1047    1.4603   18.1049   20.4621
       0.2094    1.4917   17.4000   21.2342
       0.3142    1.5440   16.6373   22.1089
       0.4189    1.6171   15.8717   23.0315
       0.5236    1.7109   15.1179   23.9872
       0.6283    1.8253   14.3814   24.9709
       0.7330    1.9600   13.6649   25.9800
       0.8378    2.1147   12.9698   27.0133
       0.9425    2.2890   12.2974   28.0700
       1.0472    2.4824   11.6486   29.1497
       1.1519    2.6940   11.0245   30.2523
       1.2566    2.9226   10.4265   31.3775
       1.3614    3.1666    9.8564   32.5252
       1.4761    3.4233    9.3169   33.6951
       1.5708    3.6884    8.8125   34.8874
       1.6755    3.9548    8.3501   36.1018
       1.7802    4.2107    7.9419   37.3383
       1.8850    4.4355    7.6082   38.5968
       1.9897    4.5970    7.3814   39.8768
       2.0944    4.6571    7.2995   41.1096


DO YOU WANT TO PLOT THESE RESULTS
TYPE YES OR NO
YES
```

E-K DIAGRAM SHOWING FIRST THREE ENERGY LEVELS

```
DO YOU WANT TO CALCULATE THE EFFECTIVE MASS
 TYPE YES OR NO
YES


THE MASS OF A FREE ELECTRON IS 1.0
```

K	M1		M2		M3	
0.0000	1.05	E	0.03	H	0.03	E
0.1047	1.05	E	0.07	H	0.06	E
0.2094	1.05	E	0.38	H	0.21	E
0.3142	1.05	E	7.40	H	0.46	E
0.4189	1.06	E	1.86	E	0.66	E
0.5236	1.07	E	1.27	E	0.79	E
0.6283	1.08	E	1.10	E	0.86	E
0.7330	1.10	E	1.02	E	0.91	E
0.8378	1.12	E	0.97	E	0.93	E
0.9425	1.15	E	0.93	E	0.96	E
1.0472	1.20	E	0.89	E	0.96	E
1.1519	1.28	E	0.84	E	0.97	E
1.2566	1.43	E	0.79	E	0.98	E
1.3614	1.73	E	0.71	E	0.98	E
1.4661	2.60	E	0.63	E	0.99	E
1.5708	15.80	E	0.52	E	0.99	E
1.6755	2.07	H	0.41	E	0.99	E
1.7802	0.71	H	0.29	E	1.00	E
1.8850	0.35	H	0.21	E	1.02	E
1.9897	0.22	H	0.15	E	0.47	H
2.0944	0.18	H	0.13	E	0.01	H

```
DO YOU WANT TO PLOT THESE RESULTS
 TYPE YES OR NO
NO
```

```
DO YOU WANT TO CALCULATE THE PROBABILITY DENSITY
 FUNCTIONS FOR THE FIRST THREE ENERGY STATES
 TYPE YES OR NO
YES
 INPUT A VALUE OF K AS A REAL NUMBER
2.0944
```

X	PD1	PD2	PD3	V
-0.7500	-0.0000	1.7479	-0.0000	5.0000
-0.6750	0.0410	1.7247	0.3807	5.0000
-0.6000	0.1640	1.6579	1.2313	5.0000
-0.5250	0.3712	1.5536	1.9127	5.0000
-0.4500	0.6541	1.3909	1.8729	0.0000
-0.3750	0.9788	1.1409	1.0989	0.0000
-0.3000	1.3072	0.8310	0.2388	0.0000
-0.2250	1.6075	0.5166	0.0327	0.0000
-0.1500	1.8471	0.2462	0.6556	0.0000
-0.0750	2.0014	0.0641	1.5750	0.0000
0.0000	2.0552	-0.0000	2.0028	0.0000
0.0750	2.0014	0.0641	1.5750	0.0000
0.1500	1.8471	0.2462	0.6556	0.0000
0.2250	1.6075	0.5166	0.0327	0.0000
0.3000	1.3072	0.8310	0.2388	0.0000
0.3750	0.9788	1.1409	1.0989	0.0000
0.4500	0.6541	1.3909	1.8729	0.0000
0.5250	0.3712	1.5536	1.9127	5.0000
0.6000	0.1640	1.6579	1.2313	5.0000
0.6750	0.0410	1.7247	0.3807	5.0000
0.7500	-0.0000	1.7479	-0.0000	5.0000

```
DO YOU WANT TO PLOT THESE RESULTS
 TYPE YES OR NO
YES
```

PROBABILITY DENSITY VS X FOR FIRST THREE STATES.
THE POTENTIAL IS SHOWN AS 4

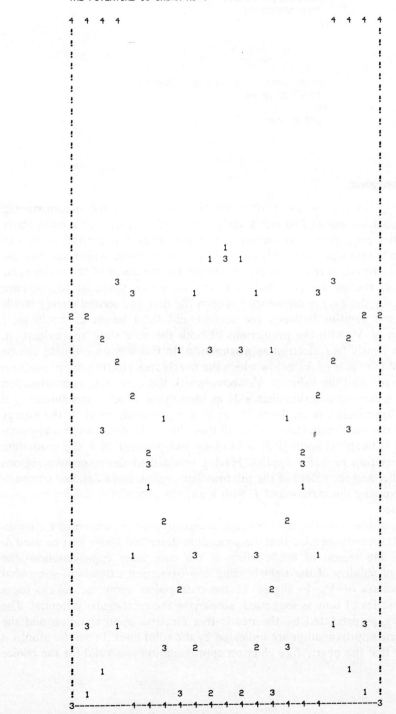

```
DO YOU WANT TO CALCULATE ANOTHER SET OF
PROBABILITY DENSITY FUNCTIONS
TYPE YES OR NO
NO



DO YOU WANT TO TRY ANOTHER POTENTIAL
TYPE YES OR NO
NO
END OF RUN
```

3.2 Discussion

The program can be used in a variety of ways and the accompanying documentation tailored to suit a variety of levels of user, but a particularly instructive procedure is to concentrate on one form of potential and first to perform a series of runs, varying V_0 over a wide range whilst keeping the period constant. It is then relatively simple for the cases of the rectangular potential, the cosine potential, and the harmonic potential to compare graphically the energy difference between the first and second energy bands at $k = \pi/a$ (and/or between the second and third bands at $k = 0$) as a function of V_0 with the predictions of both the core state approximation, and the nearly-free electron approximation. In this way an estimate can be made of the value of V_0 below which the nearly free electron approximation is accurate, and the value of V_0 above which the core state approximation holds. A second criterion that aids in identifying which approximation, if any, is appropriate is the form of the $E-k$ relationship itself. If the energy gaps at the zone boundaries are small then the nearly-free electron approximation is likely to hold; if E is virtually independent of k the core state approximation probably applies. Having established the respective regions of validity and the extent of the intermediate region, more detailed comparisons involving the variation of E with k and the probability density can then be made.

Comparison with the tight-binding approximation is rather more complicated. It is recommended that the procedure described above first be used to identify the region of applicability of the core state approximation; the region of validity of the tight-binding approximation extends to somewhat lower values of V_0. In Figure 12 the data points show the results for a typical series of runs as suggested, employing the rectangular potential. The energy gaps predicted by the nearly-free electron approximation and the core state approximation are indicated by the solid lines. From the graphs it is clear that the nearly-free electron approximation was valid for the choice

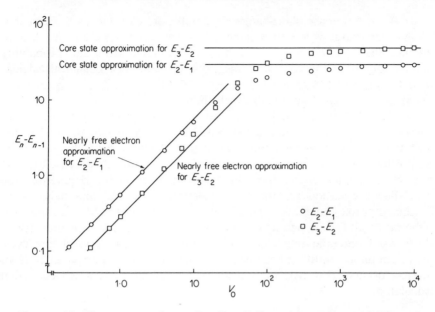

Figure 12. The computed results for $E_2(k = \pi/a) - E_1(k = \pi/a)(\bigcirc)$ and $E_3(k = 0) - E_2(k = 0)(\square)$ as a function of V_0 for a rectangular potential of period 1.5 and with a rectangle width (b) of 0.5. The solid lines show the predictions of the nearly-free electron approximation (for low V_0) and the core state approximation (for high V_0)

of parameters employed in section 3.2. This conclusion is consistent with the variation of E with k and the form of the probability density found in that case.

4. FURTHER CALCULATIONS

4.1 Estimating the potential

In the present program the potential can be varied widely in form, magnitude, and period. In a full-scale calculation for a particular material it is chosen to conform as closely as possible to the actual potential. It should, however, be noted that treating the problem as that of finding the eigenstates of a single electron in a potential due to the nuclei and the other electrons is itself an approximation. In principle the true wave function should embrace all the interacting particles. Fortunately, it appears that the results for the important physical properties are generally satisfactory using the single-electron approximation.

The principal contributions to the potential are taken to be:

(i) A long-range Coulomb potential due to each of the atomic nuclei screened by their associated core state electrons less the band electrons.

(ii) A more intense short-range Coulomb potential due to the fact that the nuclear charge is less fully screened by the core state electrons close to the nucleus. The degree of screening is estimated from the probability density found for the core state electrons in quantum-mechanical calculations for the isolated atom.

(iii) An 'exchange' interaction with the core state electrons.

(iv) A Coulomb interaction with all of the *other* band electrons. This contribution to the potential is estimated by assuming that the interaction with the band electrons is equivalent to that produced by a charge density proportional to the sum of the band electron probability densities. The procedure adopted is iterative; on the basis of an assumed potential the Schrödinger equation for the band electrons is solved and hence their combined probability density estimated. This leads in general to a somewhat different result for the potential than was assumed originally. On the basis of the revised potential the Schrödinger equation is again solved, the cycle being continued until no significant change in any of the parameters of interest occurs. This type of procedure is referred to as a self-consistent calculation.

4.2 The three-dimensional case

In a normal crystalline conductor the band electrons are free to move in three dimensions. The major difference introduced by this elaboration into the formalism of band calculations is that several quantities which can be treated as scalars in a one-dimensional model have to be recognized as vectors in the three-dimensional case. This is obviously the case for the variable used to represent position in the crystal; x has been used here for the one-dimensional case, $\mathbf{r}$ is generally used in the three-dimensional case. Similarly, the scalar k becomes the vector $\mathbf{k}$. The one-dimensional free electron wave function $\psi(x) = \exp(ikx)$ becomes $\psi(\mathbf{r}) = \exp(i\mathbf{k} \cdot \mathbf{r})$ and specifies the value at $\mathbf{r}$ of a wave function of wavelength $\lambda = 2\pi/|\mathbf{k}|$ propagating in the direction of the vector $\mathbf{k}$ (the wave vector). One speaks of '$\mathbf{k}$-space' or 'reciprocal space' ($\mathbf{k}$ has units L^{-1}).

The quantity corresponding to $2n\pi/a$, which arises as a consequence of Bragg reflections and determines the zone boundaries, must also be expressed as a vector in the three-dimensional case. The condition for Bragg reflection is $2a \sin \theta = n\lambda$ (where a is the spacing between the atomic *planes*), i.e.

$$2a \sin \theta = n \frac{2\pi}{|\mathbf{k}|},$$

so that

$$|\mathbf{k}| \sin \theta = \frac{n\pi}{a}.$$

Bragg reflection therefore occurs when the component of **k** in a direction normal to a lattice plane is $n\pi/a$. The change in the component of **k** normal to the plane is $2n\pi/a$. Vectors **K**$_n$, of magnitude $2n\pi/a$ directed normally to the various lattice planes, of interplane spacing a, fill, in the three-dimensional case, the role occupied by $2n\pi/a$ in the one-dimensional case. These vectors **K**$_n$ are referred to as reciprocal lattice vectors.

The most important new feature which occurs as a consequence of these considerations in the three-dimensional case is that the electron energy now depends not only on the magnitude but also on the direction of **k**. In consequence the effective mass is likewise dependent on direction and is expressed as a second-rank tensor.

4.3 Other basis functions

Since any realistic potential involves Coulomb interactions with the atomic nuclei a large number of terms are required to represent it accurately by a Fourier series of the kind used in the program presented here. Since the algorithm used to find the eigenvalues takes a computing time approximately proportional to the cube of the number of terms employed, this feature can render the accurate solution of the problem completely impracticable. For this reason alternative basis functions are generally used in full-scale band structure calculation. Three of the most widely used approaches are as follows.

(i) The Orthogonalized Plane Wave (OPW) method[11]

The core state electrons are of no real interest in a band structure calculation—their energies and wave functions are already known from calculations on the free atom and are unchanged in the solid. In the method of orthogonalized plane waves the core states are deliberately ignored by using the Schmidt orthogonalization procedure to construct basis functions which consist of a linear combination of plane waves and core states. The basis functions so formed are orthogonal to the core states.

(ii) The Augmented Plane Wave (APW) method[12]

The potential close to an atomic nucleus in a solid is nearly spherically symmetrical about the nucleus; well away from the nuclei it is nearly constant. In the augmented plane wave approach the radial part of the Schrödinger equation in the regions close to the nuclei is solved by numerical integration. The full solutions—the product of the radial solutions and the spherical harmonics—are matched at the surface of non-overlapping spheres, centred on each of the nuclei, to plane waves (plane waves being the solutions to the Schrödinger equation in a region of constant potential).

A set of plane waves, each differing in wave vector by a reciprocal lattice vector and each matched to the solutions close to the nucleus, are used as basis functions for the variational method.

(iii) *The scattering or Korringa, Kohn, and Rostoker (KKR) method*[13]

This method is allied to the phase-shift method used in scattering theory. The basis functions are made up of ingoing and outgoing spherical waves centred on the nuclei. In the most common form of this method the Green's function approach is employed to cast the Schrödinger equation in integral form.

For materials with $Z \geqslant 55$ relativistic corrections become important, and all of these methods can, if necessary, be modified to incorporate these effects.[14–16]

In principle, provided the basis functions form a complete set, the results of a band structure calculation should be independent of the choice of basis functions—provided enough are used. In practice there are always some deviations, though they are usually small.

4.4 Metals and semiconductors

The methods employed in band structure calculations for metals and semiconductors are quite similar; their different physical properties arise because a metal has one or more partially filled bands, a semiconductor at the absolute zero of temperature has energy bands that are either completely full (the valence band and the core states) or completely empty (the conduction band).

In the case of a metal, band structure calculations lead to an estimate of the density of states, the Fermi energy, and hence of the form of the Fermi surface—the energy surface in **k**-space which delineates the boundary between the filled and the empty electron states at the absolute zero. In a metal it is the electrons close to the Fermi surface that are responsible for the electronic transport properties.

In the case of a semiconductor the most important features determined by a band structure calculation are the energy gap between the valence band and the conduction band, the effective masses and the densities of states of the hole states close to the top of the valence band, and the electron states at the bottom of the conduction band. It is these states which lead to electronic conduction in semiconductors.

REFERENCES

1. A clear discussion of band structure calculations is given by L. Pincherle, *Electronic Energy Bands in Solids* (Macdonald, London, 1971).

2. R. de L. Kronig and W. G. Penney, *Proc. Roy. Soc. (London)*, **A130,** 499 (1931).
3. H. G. Schuster (Ed.) *One Dimensional Conductors*, (Springer-Verlag, New York, 1975).
4. F. Bloch, *Zeit. Phys.*, **52,** 555 (1928).
5. R. M. Eisberg, *Fundamentals of Modern Physics* (Wiley, New York, 1961).
6. G. C. Wetsel, Jr., *Am. J. Phys.*, **46,** 714 (1978).
7. M. L. Cohen, V. Heine, and D. Weaire, *Solid State Physics*, **24,** (1970).
8. R. Barrie, *Proc. Phys. Soc.*, **69,** 553 (1956).
9. T. C. Harman and J. M. Honig, *J. Phys. Chem. Solids*, **23,** 913 (1962).
10. NAG—Numerical Algorithms Group; Central Office: Oxford University Computing Laboratory, 13, Banbury Road, Oxford OX2 6NN, U.K.
11. T. O. Woodruff, *Solid State Physics*, **4,** 367 (1957).
12. T. L. Loucks, *Augmented Plane Wave Method* (Benjamin, New York, 1967).
13. G. C. Fletcher, *The Electron Band Theory of Solids*. Ch. 12 (North-Holland, London, 1971).
14. P. Soven, *Phys. Rev.*, **A137,** 1706 (1965).
15. T. L. Loukes, *Phys. Rev.*, **A139,** 1333 (1965).
16. Y. Onodera, M. Okazaki, and T. Inui, *J. Phys. Soc. Japan*, **21,** 2173 (1966).

ENERGY BANDS

```
C     MASTER SEGMENT OF PROGRAM.
C     THIS SEGMENT CONTAINS SECTIONS WHERE DATA
C     IS INPUT FROM A TERMINAL.
C     CHANGES WOULD BE NECESSARY TO RUN THE PROGRAM
C     IN BATCH MODE.
      DOUBLE PRECISION H(33,33),A(33),E(33),EF(33,33),W(33)
      COMMON AMP,WIDTH,PERIOD,PI,NPOT,FVAL(20),XVAL(20),NVAL
      REAL K,EL(21,3),KX(21),EM(21,3),ELEC(21,3),WF(21,4),XX(21)
      LOGICAL LOG,LOG1,LOG2,LOG3
      DATA YY,YN,EE,EH/4HYES ,4HNO  ,1HE,1HH/
      PI = 4.0*ATAN(1.0)
      WRITE(1,190)
      WRITE(1,191)
      WRITE(1,192)
      WRITE(1,193)
      WRITE(1,194)
    1 WRITE(1,200)
      WRITE(1,201)
      WRITE(1,202)
      WRITE(1,203)
      WRITE(1,204)
      WRITE(1,205)
      READ(1,100)NPOT
      IF(NPOT.EQ.1) GOTO 10
      IF(NPOT.EQ.2) GOTO 20
      IF(NPOT.EQ.3) GOTO 30
      IF(NPOT.EQ.4) GOTO 40
      IF(NPOT.EQ.5) GOTO 50
      WRITE(1,207)
      GOTO 1
C       DATA FOR RECTANGULAR POTENTIAL
   10 WRITE(1,210)
      WRITE(1,211)
      READ(1,101) AMP
      WRITE(1,212)
      READ(1,101) WIDTH
      WRITE(1,208)
      READ(1,101) PERIOD
      IF(PERIOD.GT.WIDTH) GOTO 70
      WRITE(1,213)
      GOTO 10
C       DATA FOR SAWTOOTH POTENTIAL
   20 WRITE(1,220)
      WRITE(1,221)
      READ(1,101)AMP
      WRITE(1,208)
      READ(1,101)PERIOD
      GOTO 70
C       DATA FOR COSINE POTENTIAL
   30 WRITE(1,230)
      WRITE(1,231)
      WRITE(1,241)
      READ(1,101)AMP
      WRITE(1,208)
      READ(1,101)PERIOD
      GOTO 70
```

ENERGY BANDS

```
C          DATA FOR HARMONIC POTENTIAL
    40 WRITE(1,240)
       WRITE(1,241)
       READ(1,101)AMP
       WRITE(1,208)
       READ(1,101)PERIOD
       AMP=AMP*4.0/(PERIOD*PERIOD)
       GOTO 70
C          DATA FOR INTERPOLATED POTENTIAL
    50 WRITE(1,250)
       WRITE(1,251)
       WRITE(1,252)
       WRITE(1,253)
    51 WRITE(1,271)
       WRITE(1,256)
       READ(1,102)NVAL
       IF(NVAL.LT.2) GOTO 51
       IF(NVAL.GT.20) GOTO 51
    52 WRITE(1,257)
       READ(1,101)FVAL(1)
       XVAL(1)=0.0
       DO 54 I=2,NVAL
       WRITE(1,258)
       READ(1,101)XVAL(I)
       IF(XVAL(I).GT.XVAL(I-1)) GOTO 53
       WRITE(1,270)
       GOTO 52
    53 WRITE(1,259)
       READ(1,101)FVAL(I)
    54 CONTINUE
       PERIOD=2.0*XVAL(NVAL)
       GOTO 70
    70 NPTS=100.0*PERIOD
C          CALCULATION OF RESULTS FOR E-K DIAGRAM
       CALL FRANCS(33,NPTS,DC,A)
       WRITE(1,300)PERIOD
       WRITE(1,301)
       DO 2 L=1,21
       K=0.05*PI*FLOAT(L-1)/PERIOD
       LOG=.FALSE.
       IF(L.LE.10)NAR=10
       IF(L.GT.10)NAR=11
    12 CONTINUE
       CALL ENERGY(K,E,EF,NAR,NAR,DC,A,H,W,.TRUE.)
       IF(LOG)GOTO 13
       E1=E(1)
       E2=E(2)
       E3=E(3)
       NAR=NAR+2
       LOG=.TRUE.
       GOTO 12
    13 CONTINUE
       F1=E(1)
       F2=E(2)
       F3=E(3)
       LOG=.FALSE.
```

ENERGY BANDS

```
      DE=F3-F1
      IF(DE.EQ.0.0)DE=F3
      NAR=NAR+8
      LOG1=.TRUE.
      LOG2=.TRUE.
      LOG3=.TRUE.
      ERES=ABS((F1-E1)/DE)
      IF(ERES.GT.3.0E-3)LOG1=.FALSE.
      ERES=ABS((F2-E2)/DE)
      IF(ERES.GT.3.0E-?)LOG2=.FALSE.
      ERES=ABS((F3-E3)/DE)
      IF(ERES.GT.3.0E-3)LOG3=.FALSE.
      IF(LOG1.AND.LOG2.AND.LOG3)GOTO 16
      IF(NAR.LE.31)GOTO 12
C   OUTPUT WARNING IF ONE OF THE ACCURACIES IS BELOW LIMITS
      IF(.NOT.LOG1.AND.LOG2.AND.LOG3)WRITE(1,400)
 400 FORMAT(41H *** WARNING ACCURACY OF E1 ON NEXT LINE ,
    1 13HUNCERTAIN ***)
      IF(LOG1.AND.(.NOT.LOG2).AND.LOG3)WRITE(1,401)
 401 FORMAT(41H *** WARNING ACCURACY OF E2 ON NEXT LINE ,
    1 13HUNCERTAIN ***)
      IF(LOG1.AND.LOG2.AND.(.NOT.LOG3))WRITE(1,402)
 402 FORMAT(41H *** WARNING ACCURACY OF E3 ON NEXT LINE ,
    1 13HUNCERTAIN ***)
      IF((.NOT.LOG1).AND.(.NOT.LOG2).AND.LOG3)WRITE(1,403)
 403 FORMAT(48H *** WARNING ACCURACY OF E1 AND E2 ON NEXT LINE ,
    1 13HUNCERTAIN ***)
      IF((.NOT.LOG1).AND.LOG2.AND.(.NOT.LOG3))WRITE(1,404)
 404 FORMAT(48H *** WARNING ACCURACY OF E1 AND E3 ON NEXT LINE ,
    1 13HUNCERTAIN ***)
      IF(LOG1.AND.(.NOT.LOG2).AND.(.NOT.LOG3))WRITE(1,405)
 405 FORMAT(48H *** WARNING ACCURACY OF E2 AND E3 ON NEXT LINE ,
    1 13HUNCERTAIN ***)
      IF((.NOT.LOG1).AND.(.NOT.LOG2).AND.(.NOT.LOG3))WRITE(1,406)
 406 FORMAT(52H *** WARNING ACCURACY OF E1, E2 AND E3 ON NEXT LINE ,
    1 13UNCERTAIN ***)
  16 CONTINUE
      KX(L)=K
      DO 3 N=1,3
      EL(L,N)=F(N)
   3 CONTINUE
      WRITE(1,302) KX(L),(EL(L,N),N=1,3)
   2 CONTINUE
   4 WRITE(1,280)
      WRITE(1,291)
      READ(1,103) YA
      IF(YA.EQ.YN) GOTO 5
      IF(YA.NE.YY) GOTO 4
      WRITE(1,303)
      CALL GRAPH(EL,21,3,KX)
   5 WRITE(1,281)
      WRITE(1,291)
      READ(1,103) YA
      IF(YA.EQ.YN) GOTO 71
      IF(YA.NE.YY) GOTO 5
      WRITE(1,304)
```

ENERGY BANDS

```
      WRITE(1,305)
C     APPROXIMATE CALCULATION OF EFFECTIVE MASS
C     THE VALUE IS TRUNCATED AT 10**6 TIMES THE
C     MASS OF A FREE ELECTRON
      DK2=(KX(2)-KX(1))**2
      DO 7 N=1,3
      EM(1,N)=2.0*(EL(2,N)-EL(1,N))/DK2
      DO 6 L=2,20
      EM(L,N)=(EL(L+1,N)+EL(L-1,N)-2.0*EL(L,N))/DK2
    6 CONTINUE
      EM(21,N)=2.0*(EL(20,N)-EL(21,N))/DK2
    7 CONTINUE
      DO 8 L=1,21
      DO 9 N=1,3
      IF(EM(L,N).GE.0.0) ELEC(L,N)=EE
      IF(EM(L,N).LT.0.0) ELEC(L,N)=EH
      IF(ABS(EM(L,N)).GT.2.0E-6) GOTO 80
      EM(L,N)=2.0E-6
   80 EM(L,N)=2.0/ABS(EM(L,N))
    9 CONTINUE
      WRITE(1,310) KX(L),(EM(L,N),ELEC(L,N),N=1,3)
    8 CONTINUE
   81 WRITE(1,280)
      WRITE(1,291)
      READ(1,103) YA
      IF(YA.EQ.YN) GOTO 71
      IF(YA.NE.YY) GOTO 81
      WRITE(1,306)
      DO 82 L=1,21
      DO 82 N=1,3
      IF(EM(L,N).GT.50.0) EM(L,N)=50.0
   82 CONTINUE
      CALL GRAPH(EM,21,3,KX)
   71 WRITE(1,282)
      WRITE(1,283)
      WRITE(1,291)
      READ(1,103) YA
      IF(YA.EQ.YN) GOTO 78
      IF(YA.NE.YY) GOTO 71
   72 WRITE(1,284)
      READ(1,101) K
C     CALCULATION OF PROBABILITY DENSITY FUNCTIONS
      CALL ENERGY(K,E,EF,15,15,DC,A,H,W,.FALSE.)
      DO 73 I=1,21
      XX(I)=0.05*FLOAT(I-11)*PERIOD
      X=XX(I)
      WF(I,4)=V(X)
      DO 74 J=1,3
      WF(I,J)=PROB(J,EF,15,15,X)
   74 CONTINUE
   73 CONTINUE
      WRITE(1,307)
      DO 75 I=1,21
      WRITE(1,302) XX(I),(WF(I,J),J=1,4)
   75 CONTINUE
   76 WRITE(1,280)
```

ENERGY BANDS

```
      WRITE(1,291)
      READ(1,103) YA
      IF(YA.EQ.YN) GOTO 77
      IF(YA.NE.YY) GOTO 76
      WRITE(1,308)
      WRITE(1,309)
      PDMAX=WF(1,1)
      DO 771 I=1,21
      DO 771 J=1,3
      IF(WF(I,J).GT.PDMAX) PDMAX=WF(I,J)
  771 CONTINUE
      VMAX=ABS(WF(1,4))
      DO 772 I=2,21
      IF(ABS(WF(I,4)).GT.VMAX) VMAX=ABS(WF(I,4))
  772 CONTINUE
      IF(VMAX.LT.0.1*PDMAX) VMAX=0.5*PDMAX
      SCALE=1.5*PDMAX/VMAX
      DO 773 I=1,21
      WF(I,4)=WF(I,4)*SCALE
  773 CONTINUE
      CALL GRAPH(WF,21,4,XX)
   77 WRITE(1,285)
      WRITE(1,286)
      WRITE(1,291)
      READ(1,103) YA
      IF(YA.EQ.YN) GOTO 78
      IF(YA.NE.YY) GOTO 77
      GOTO 72
   78 WRITE(1,290)
      WRITE(1,291)
      READ(1,103) YA
      IF(YA.EQ.YN) GOTO 90
      IF(YA.NE.YY) GOTO 78
      GOTO 1
   90 WRITE(1,299)
      STOP
C     FORMATS FOR INTERACTIVE SECTIONS OF PROGRAM.
  100 FORMAT(I1)
  101 FORMAT(F10.4)
  102 FORMAT(I2)
  103 FORMAT(A4)
  190 FORMAT(1X,49HPROGRAM TO CALCULATE FIRST THREE ENERGY LEVELS OF)
  191 FORMAT(1X,50HAN ELECTRON SUBJECT TO A GIVEN PERIODIC POTENTIAL.)
  192 FORMAT(1X,39HALL INPUT AND OUTPUT IS IN ATOMIC UNITS)
  193 FORMAT(1X,44HI.E. THE UNIT OF DISTANCE IS ONE BOHR RADIUS)
  194 FORMAT(1X,33HTHE UNIT OF ENERGY IS ONE RYDBERG//)
  200 FORMAT(1X,34HWHAT SORT OF POTENTIAL DO YOU WANT)
  201 FORMAT(1X,32HTYPE 1 FOR RECTANGULAR POTENTIAL)
  202 FORMAT(1X,29HTYPE 2 FOR SAWTOOTH POTENTIAL)
  203 FORMAT(1X,27HTYPE 3 FOR COSINE POTENTIAL)
  204 FORMAT(1X,29HTYPE 4 FOR HARMONIC POTENTIAL)
  205 FORMAT(1X,33HTYPE 5 FOR INTERPOLATED POTENTIAL)
  207 FORMAT(1X,36HNUMBER INPUT MUST BE BETWEEN 1 AND 5)
  208 FORMAT(1X,27HINPUT PERIOD AS PEAL NUMBER)
  210 FORMAT(1X,21HRECTANGULAR POTENTIAL)
  211 FORMAT(1X,40HINPUT HEIGHT OF RECTANGLE AS REAL NUMBER)
```

ENERGY BANDS

```
212 FORMAT(1X,39HINPUT WIDTH OF RECTANGLE AS REAL NUMBER)
213 FORMAT(1X,46HPERIOD MUST BE GREATER THAN WIDTH OF RECTANGLE)
220 FORMAT(1X,18HSAWTOOTH POTENTIAL)
221 FORMAT(1X,39HINPUT HEIGHT OF SAWTOOTH AS REAL NUMBER)
230 FORMAT(1X,37HCOSINE POTENTIAL I.E. PROPORTIONAL TO)
231 FORMAT(1X,24H1.0-COS(2.0*PI*X/PERIOD))
240 FORMAT(1X,49HHARMONIC POTENTIAL I.E. PROPORTIONAL TO X SQUARED)
241 FORMAT(1X,48HINPUT CONSTANT OF PROPORTIONALITY AS REAL NUMBER)
250 FORMAT(1X,22HINTERPOLATED POTENTIAL)
251 FORMAT(1X,46HTHE POTENTIAL IS GIVEN BY LINEAR INTERPOLATION)
252 FORMAT(1X,44HBETWEEN POINTS SPECIFIED OVER HALF A PERIOD.)
253 FORMAT(1X,38HTHE PERIOD IS TWICE THE FINAL X VALUE.)
256 FORMAT(1X,22HINPUT NUMBER OF POINTS)
257 FORMAT(1X,43HINPUT AS REAL NUMBER THE POTENTIAL AT X=0.0)
258 FORMAT(1X,33HINPUT NEXT X VALUE AS REAL NUMBER)
259 FORMAT(1X,30HINPUT POTENTIAL AS REAL NUMBER)
270 FORMAT(1X,42HEACH X VALUE MUST BE GREATEP THAN LAST ONE)
271 FORMAT(1X,46HTHE NUMBER OF POINTS MUST BE BETWEEN 02 AND 20)
280 FORMAT(//33HDO YOU WANT TO PLOT THESE RESULTS)
281 FORMAT(//43HDO YOU WANT TO CALCULATE THE EFFECTIVE MASS)
282 FORMAT(//48HDO YOU WANT TO CALCULATE THE PROBABILITY DENSITY)
283 FORMAT(1X,43HFUNCTIONS FOR THE FIRST THPEE ENERGY STATES)
284 FORMAT(1X,35HINPUT A VALUE OF K AS A REAL NUMBER)
285 FORMAT(1X,39HDO YOU WANT TO CALCULATE ANOTHER SET OF)
286 FORMAT(1X,29HPROBABILITY DENSITY FUNCTIONS)
290 FORMAT(////36HDO YOU WANT TO TRY ANOTHER POTENTIAL)
291 FORMAT(1X,14HTYPE YES OR NO)
299 FORMAT(1X,10HEND OF RUN)
C     ********************************************************
300 FORMAT(//6X,6HPERIOD,F12.1)
301 FORMAT(/7X,1HK,9X,2HE1,8X,2HE2,8X,2HE3)
302 FORMAT(1X,5F10.4)
303 FORMAT(//5X,45HE-K DIAGRAM SHOWING FIRST THREE ENERGY LEVELS//)
304 FORMAT(//34HTHE MASS OF A FREE ELECTRON IS 1.0)
305 FORMAT(//7X,1HK,10X,2HM1,9X,2HM2,9X,2HM3)
306 FORMAT(//5X,45HABS(EFFECTIVE MASS) VS K IN FIRST THREE BANDS)
307 FORMAT(//7X,1HX,9X,3HPD1,7X,3HPD2,7X,3HPD3,7X,1HV)
308 FORMAT(//5X,48HPROBABILITY DENSITY VS X FOR FIRST THREE STATES.)
309 FORMAT(1X,27HTHE POTENTIAL IS SHOWN AS 4//)
?10 FORMAT(1X,F10.4,3(F9.2,1X,A1))
    END
C
    SUBROUTINE ENERGY(K,E,EV,ND,N,DC,A,H,W,BOOL)
C
C     SUBROUTINE TO CALCULATE THE FIRST N ENERGY LEVELS
C·    AND CORRESPONDING STATE VECTORS FOR GIVEN K AND
C     GIVEN PERIODIC POTENTIAL.
C     F02ABF IS A NAG LIBRARY ROUTINE THAT CALCULATES
C     THE EIGENVALUES AND EIGENVECTORS OF A REAL
C     SYMMETRIC MATRIX.
C     F02AAF IS A NAG LIBRARY ROUTINE THAT CALCULATES
C     JUST THE EIGENVALUES OF A REAL SYMMETRIC MATRIX.
C
    DOUBLE PRECISION E(N),H(ND,N),A(N),EV(ND,N),W(N)
    LOGICAL BOOL
    COMMON AMP,WIDTH,PERIOD,PI,NPOT,FVAL(20),XVAL(20),NVAL
```

ENERGY BANDS

```
      REAL K
      N1=N/2
      IF(2*N1.EQ.N) GOTO 1
      N2=(N+1)/2
      DO 2 I=1,N
      H(I,I)=(K-FLOAT(N2-I)*2.0*PI/PERIOD)**2+DC
    2 CONTINUE
      GOTO 3
    1 DO 4 I=1,N
      H(I,I)=(K-FLOAT(N1-I+1)*2.0*PI/PERIOD)**2+DC
    4 CONTINUE
    3 NM=N-1
      DO 5 I=1,NM
      IP=I+1
      DO 5 J=IP,N
      JMI=J-I
      H(I,J)=A(JMI)/2.0
      H(J,I)=H(I,J)
    5 CONTINUE
      IFAIL=0
      IF(BOOL)CALL F02AAF(H,ND,N,E,W,IFAIL)
      IF(.NOT.BOOL)CALL F02ABF(H,ND,N,E,EV,ND,W,IFAIL)
      RETURN
      END
C
      SUBROUTINE FRANCS(NTERMS,NDATA,DC,A)
C
C        SUBROUTINE TO FOURIER ANALYSE THE POTENTIAL.
C        FOR THE INTERPOLATED POTENTIAL THE COEFFICIENTS
C        ARE EVALUATED BY NUMERICAL INTEGRATION.
C        FOR THE OTHER POTENTIALS ANALYTIC EXPRESSIONS
C        FOR THE FOURIER COEFFICIENTS ARE USED.
C
      DOUBLE PRECISION A(NTERMS)
      COMMON AMP,WIDTH,PERIOD,PI,NPOT,FVAL(20),XVAL(20),NVAL
      REAL K
      IF(NPOT.EQ.1) GOTO 10
      IF(NPOT.EQ.2) GOTO 20
      IF(NPOT.EQ.3) GOTO 30
      IF(NPOT.EQ.4) GOTO 40
      GOTO 50
C        FOURIER COEFFICIENTS FOR RECTANGULAR POTENTIAL
   10 DC=WIDTH*AMP/PERIOD
      DO 11 M=1,NTERMS
      FM=M
      A(M)=2.0*COS(FM*PI)*AMP*SIN(FM*PI*WIDTH/PERIOD)/(FM*PI)
   11 CONTINUE
      RETURN
C        FOURIER COEFFICIENTS FOR SAWTOOTH POTENTIAL
   20 DC=AMP/2.0
      DO 21 M=1,NTERMS
      FM=M
      A(M)=2.0*AMP*(COS(FM*PI)-1.0)/(FM*PI)**2
   21 CONTINUE
      RETURN
C        FOURIER COEFFICIENTS FOR COSINE POTENTIAL
```

ENERGY BANDS

```
   30 DC=AMP
      A(1)=-AMP
      DO 31 M=2,NTERMS
      A(M)=0.0
   31 CONTINUE
      RETURN
C        FOURIER COEFFICIENTS FOR HARMONIC POTENTIAL
   40 DC=AMP*PERIOD**2/12.0
      DO 41 M=1,NTERMS
      FM=M
      A(M)=AMP*PERIOD**2*COS(FM*PI)/(FM*PI)**2
   41 CONTINUE
      RETURN
C        FOURIER COEFFICIENTS BY NUMERICAL INTEGRATION
   50 K=2.0*PI/PERIOD
      FD=NDATA
      STEP=PERIOD/FD
      DC=V(0.0)
      DO 51 M=1,NTERMS
      A(M)=V(0.0)
   51 CONTINUE
      ND1=NDATA-1
      DO 52 I=1,ND1
      FI=I
      X=FI*STEP
      VX=V(X)
      DC=DC+VX
      DO 53 M=1,NTERMS
      FM=M
      A(M)=A(M)+COS(FM*K*X)*VX
   53 CONTINUE
   52 CONTINUE
      DC=DC/FD
      DO 54 M=1,NTERMS
      A(M)=A(M)*2.0/FD
   54 CONTINUE
      RETURN
      END
C
      FUNCTION V(Y)
C
C        FUNCTION TO EVALUATE THE POTENTIAL AT ANY POINT.
C        THE POTENTIAL IS ALWAYS AN EVEN FUNCTION.
C
      COMMON AMP,WIDTH,PERIOD,PI,NPOT,FVAL(20),XVAL(20),NVAL
      X=Y
      IF(X.LT.0.0) X=-X
      P2=PERIOD/2.0
      R=AMOD(X,PERIOD)
      IF(R.GT.P2) R=PERIOD-R
      IF(NPOT.EQ.1) GOTO 10
      IF(NPOT.EQ.2) GOTO 20
      IF(NPOT.EQ.3) GOTO 30
      IF(NPOT.EQ.4) GOTO 40
      GOTO 50
C        RECTANGULAR POTENTIAL
```

ENERGY BANDS

```
   10 B=(PERIOD-WIDTH)/2.0
      IF(R.LT.B) V=0.0
      IF(R.EQ.B) V=AMP/2.0
      IF(R.GT.B) V=AMP
      RETURN
C        SAWTOOTH POTENTIAL
   20 V=R*AMP/P2
      RETURN
C        COSINE POTENTIAL
   30 Q=2.0*PI/PERIOD
      V=AMP*(1.0-COS(Q*R))
      RETURN
C        HARMONIC POTENTIAL
   40 V=AMP*R**2
      RETURN
C        INTERPOLATED POTENTIAL
   50 NV1=NVAL-1
      DO 51 I=1,NV1
      IF(R.GE.XVAL(I).AND.R.LT.XVAL(I+1)) J=I
   51 CONTINUE
      GRAD=(FVAL(J+1)-FVAL(J))/(XVAL(J+1)-XVAL(J))
      V=FVAL(J)+GRAD*(R-XVAL(J))
      RETURN
      END
C
      FUNCTION PROB(L,EV,ND,N,X)
C
C        FUNCTION TO EVALUATE THE PROBABILITY DENSITY AT X
C        FOR GIVEN K AND GIVEN STATE VECTOR.
C
      DOUBLE PRECISION EV(ND,N)
      COMMON AMP,WIDTH,PERIOD,PI,NPOT,FVAL(20),XVAL(20),NVAL
      S=0.0
      ND1=ND-1
      DO 1 I=1,ND1
      I1=I+1
      DO 2 J=I1,ND
      F=2.0*PI*FLOAT(J-I)/PERIOD
      S=S+EV(I,L)*EV(J,L)*COS(F*X)
    2 CONTINUE
    1 CONTINUE
      PROB=1.0+2.0*S
      RETURN
      END
C
      SUBROUTINE GRAPH(Y,NP,NG,X)
C
C        SUBROUTINE TO PLOT UP TO FOUR FUCTIONS
C        ON THE SAME GRAPH.
C
      DIMENSION Y(NP,NG),X(NP),PLOT(120),PLT(7)
      DATA PLT(1),PLT(2),PLT(3),PLT(4),PLT(5),PLT(6),PLT(7)/
     1 4H1   ,4H2   ,4H3   ,4H4   ,4H    ,4H-   ,4H!   /
      IWG=70
      GAP=FLOAT(IWG-1)/FLOAT(NP-1)
      IGAP=IFIX(AINT(GAP))
```

ENERGY BANDS

```
      IWG=(NP-1)*IGAP+1
      YMIN=Y(1,1)
      YMAX=Y(1,1)
      DO 66 N=1,NG
      DO 68 I=1,NP
      IF(Y(J,N).LT.YMIN) YMIN=Y(I,N)
      IF(Y(I,N).GT.YMAX) YMAX=Y(I,N)
   68 CONTINUE
   66 CONTINUE
      IF(YMIN.GT.0.0) YMIN=0.0
      IHG=IWG
      SCALE=(YMAX-YMIN)/FLOAT(IHG-1)
      DO 70 J=1,IHG
      YTS=YMAX-SCALE*(FLOAT(J)-1.5)
      YBS=YMAX-SCALE*(FLOAT(J)-0.5)
      PLOT(1)=PLT(7)
      PLOT(IWG)=PLT(7)
      DO 72 I=1,NP
      K1=(I-1)*IGAP+1
      K2=I*IGAP
      DO 74 K=K1,K2
      IF(K.NE.1.AND.K.NE.IWG) PLOT(K)=PLT(5)
      IF(YTS.GE.0.0.AND.YBS.LT.0.0) PLOT(K)=PLT(6)
   74 CONTINUE
      DO 73 N=1,NG
      IF(Y(I,N).LE.YTS.AND.Y(I,N).GT.YBS) PLOT(K1)=PLT(N)
   73 CONTINUE
   72 CONTINUE
      WRITE(1,178)(PLOT(K),K=1,IWG)
  178 FORMAT(1H ,120A1)
   70 CONTINUE
   62 CONTINUE
      RETURN
      END
```